Echoes of Afghanistan

THE COST OF SURVIVAL

THE CORBIN SIPE STORY

Geoffrey Yoder

Written by Geoffrey Yoder (As told to Geoffrey by Corbin Sipe)

Editors: Special thanks for the Heritage Shores Military Club members for the instructive inputs and edits

eBook ISBN: 979-8-9987404-6-6

Paperback ISBN: 979-8-9987404-5-9

Front Cover Credit: Katrina Sipe

All other photos provided by Corbin and Katrina Sipe, and Lauretta Yoder

Special thank-you to Corbin and Katrina Sipe for their courage in sharing their deeply personal journey with me, which sheds light on the resilience required to navigate the aftermath of military service. Corbin's tour of duty in Afghanistan, marked by unwavering commitment and sacrifice, left indelible scars on his psyche. Battling Post-Traumatic Stress Disorder (PTSD) and the haunting weight of survivor's guilt, he has walked a path fraught with invisible wounds—struggles that resonate beyond the battlefield. Katrina's steadfast support and advocacy alongside him underscore the strength of their partnership, transforming vulnerability into a testament of hope. By allowing their story to be told, they honor those who served alongside Corbin, while offering a reminder of the profound, often unseen challenges faced by veterans and their families. Their willingness to confront these truths sparks vital conversations, fostering empathy and understanding for all who grapple with the long shadows of war.

Contents

Dedication

From SSgt Corbin Sipe

Life's most profound lessons are often forged in struggle, shaped by the hands of those who refuse to let us surrender to the darkness. This book is not just a collection of words, but a testament to resilience, a journey made possible by the unwavering love and support of those who stood beside me when the road seemed insurmountable.

To my cherished wife, whose unwavering faith became my anchor when storms threatened to pull me under, I dedicate this work with boundless gratitude. Your strength in my weakest moments reminded me that even the fiercest battles are lighter when shared.

To Bud, whose steadfast belief in me never wavered, thank you for refusing to let me give up, for guiding me through the REBOOT program, and for proving that redemption is always within reach.

And to my brothers in the Marine Corps, whose unshakable brotherhood and trust in my abilities fortified my spirit, you taught me that true strength is not just endurance, but the courage to rise after every fall.

This book is also for every reader who stands at the edge of their own trials, wondering if perseverance is worth the pain. Let these pages be your compass, a reminder that no dream is too distant, no obstacle too great, when met with relentless determination.

Whether you are searching for hope, courage, or simply proof that resilience wins, know this: you are not alone. My journey is proof that even the most broken paths can lead to triumph, if only we refuse to stop walking.

With deepest sincerity and gratitude, I invite you to take this step forward, together.

Corbin Sipe

Praise for the Book

As I reflect on my experiences performing for our U.S. soldiers in Iraq during a two-week USO tour in February 2010, I am reminded of the profound impact that their sacrifices have had on my life. I am reminded of the vibrant young lives that were lost in the pursuit of freedom. Their laughter, their smiles, and their stories are forever etched in my memory, a bittersweet reminder of the transience of life and the permanence of sacrifice. Memories of their bravery, camaraderie, and resilience are etched in my mind like scars on a warrior's body.

But among the countless stories of valor and sacrifice I witnessed throughout my career, one stands out in my mind is the story of Marine Corps SSgt Corbin Sipe as told in this book, **Echoes of Afghanistan.** SSgt Sipe's narrative is a testament to the unwavering dedication and unrelenting courage that defines our men and women in uniform. His selfless act of service is a poignant reminder of the immense sacrifices that are made daily by those who put country before self. As I think about Corbin's story, I am filled with a sense of awe and gratitude that is almost overwhelming.

The weight of his sacrifice, and that of his fellow Marines who perished in Afghanistan, is a burden that our nation must never forget. We must honor their memory by ensuring that their sacrifices are not in vain. The families of these brave men and women, who are forever changed by the loss of their loved ones, deserve our deepest respect and support. Their pain and grief are a reminder that the cost of peace is not just paid by those who serve, but also by those who love them.

As I look back on my experiences, I am filled with a sense of purpose and responsibility. I am committed to ensuring that the sacrifices of our troops, including SSgt Sipe and his fellow Marines, are never forgotten. Their stories must continue to be told, their memories must be honored, and their families must be supported. We owe it to them, and to ourselves, to remember the immense cost of peace and the indomitable spirit of those who defend it.

In the end, Corbin's story is not just a reminder of the sacrifices made by our troops, but also a testament to the power of courage, duty, and sacrifice. His legacy, and that of his fellow Marines, must continue to inspire us to stand up for what we believe in, to fight for our freedom, and to support those who defend it. We must never forget the lessons of their bravery, and we must always honor their memory by striving to be worthy of their sacrifice. ***Kaya Jones, Singer/Performer/Grammy Award Winner***

***

I want to tell you what a privilege it was to read your book before it was published!

I thought the book was very emotional, and informational about what our military goes through that nonmilitary has no clue about. I wish it was mandatory reading for Americans, especially those that don't think our military is important or how long it affects them when their duty time is over. It was wonderful to see there are resources for those that suffer PTSD, but they need to seek help.

Thank you to Corbin for baring his soul so that it might help others, and to you for helping him and many others with writing this book. ***Kathy Harrigan, Mother of two Veterans, and wife of a Veteran***

***

I am deeply grateful for this profoundly moving book, which illuminates the hidden hardships that service members endure—experiences that most civilians can scarcely comprehend. Its raw emotion and hard-won knowledge are essential for those who underestimate the military's significance or fail to grasp the lingering effects of duty after a service member returns home. The inclusion of concrete resources for veterans coping with PTSD is both reassuring and crucial. However, we must also do our part to let them know they are not alone and that the resources are available to them. The high suicide rate among veterans underscores the urgency of our involvement now.

My deepest gratitude goes to Corbin for sharing his soul so candidly. He offers a beacon of hope for others who may be suffering in silence. To the team that helped shape his story into a book that could give the military the recognition and understanding it deserves, I extend my heartfelt thanks. ***Branden Zeitler, Delaware Local Coordinator, Irreverent Warriors Foundation***

***

This book is a powerful and moving testament to the hidden hardships faced by service members, challenges that often elude the understanding of civilians. It delves into raw emotions that are vital for anyone who might underestimate the military's significance or fail to grasp the enduring impact of duty on service members returning home. The book is not just a narrative; it is a lifeline, offering resources for veterans coping with PTSD and providing both reassurance and essential support. It is a call to action, urging us to actively communicate to veterans that they are not alone and that help is readily available, especially considering the alarming suicide rate. As Director for Veterans Outreach at People's Place, I see every day the critical importance of addressing both the physical and mental needs of veterans, from securing shelter to providing mental health support.

I am deeply grateful to Corbin for sharing his soul so candidly, becoming a beacon of hope for those who suffer in silence. This book is a must-read for anyone who wishes to truly understand and support our veterans. ***Michael Rowe, Program Director for Veterans Outreach, People's Place, Milford DE***

Forward

This is a story not easily spoken, but one that desperately needs to be heard. We embark on writing this book together, Corbin and I, not just to recount events, but to offer a space for healing, a beacon of understanding for fellow warriors, and a heartfelt embrace for every Gold Star family walking a similar path.

Corbin's youth, like many, was of uncertainty and discovery. There were moments of doubt, the kind that whispers in the quiet hours, alongside the spirited challenges that build character, one hard-won lesson at a time. These early experiences, perhaps unknowingly, laid the groundwork for the profound dedication he would later give to the United States Marine Corps. It was a commitment born of something deeper than ambition – a call to serve, to protect, to become part of something greater than himself.

The uniform brought with it a new chapter of challenges, both physical and mental, forging a spirit of unwavering bravery. He learned the meaning of brotherhood, the unspoken language between those who stand shoulder-to-shoulder, ready to face the unknown. These were the lessons that bonded him to his team, a sacred kinship that would forever define a part of him. Then came Afghanistan. The dust, the ever-present tension created bonds that were either shattered or hardened into steel. It was there, amidst the chaos and the profound acts of courage, that the deepest sorrow found its way into his life. The loss of three of his team members, his brothers, left a chasm in his heart, a sadness so profound it defies easy words.

Reliving these memories is, for Corbin, an act of immense courage. Each word spoken, each scene revisited, brings with it a fresh wave of grief, a vivid echo of what was lost. Yet he chooses to tell it. He chooses to lay bare the vulnerabilities and the enduring pain, not for pity, but for connection. He understands that these heart-wrenching stories, though difficult to tell and to hear, are vital. They are the threads that weave us together, assuring others that they are not alone in their journey of remembrance, sorrow, and resilience.

With deep appreciation for your willingness to listen, Corbin and I hope that the words within this story will resonate within you, filling your emotions with compassion for the trials faced by our service men and women. We hope it sparks an empathy that extends to Gold Star families, those who continue their journey with only memories of their fallen warriors, carrying a sacred torch of love and loss. May this narrative serve as a heartfelt charge to support all who serve and all who grieve, recognizing the immeasurable sacrifices made for us all.

Chapter One

My Interview with Corbin

The air in my office was thick with the familiar aroma of freshly brewed coffee as I started the Zoom call with Corbin Sipe. Corbin appeared on my screen; a sharp, clear image projected through the impersonal lens of technology. As we spoke, the distance seemed to melt away, replaced by a closeness forged by the gravity of the subject matter.

This call was intended to be a logistical chat, a gentle probing of sensitive boundaries before we fully embarked on the detailed documentation of the tragedy of April 8, 2019; the day that forever changed his life as a Marine in Afghanistan. As Corbin began to unfold the quiet prelude to his story, a different kind of intensity filled the air transcending the miles and the screen between us. Before the questions began, my mind reflected on our previous less formal discussions about his journey, a quiet, almost reverent anticipation building within me. An immense ache of pain and a sense of hope settled over me, a testament to the quiet power of his narrative even in its early stages.

His eyes, though holding a depth that spoke of experiences far beyond my own comprehension, were strangely kind and surprisingly gentle. They were the eyes of someone who had seen too much, the human capacity for destruction and extraordinary sacrifice, but had chosen to filter what remained through a lens of quiet resilience and compassion. His voice, steady and calm, belied the intensity of the memories he was preparing to share. As Corbin spoke about the simple preparations for our upcoming sessions, the gentle probing of what specific events might be too raw or too sensitive, I could feel the invisible weight he carried. It wasn't overt suffering, no dramatic pronouncements of trauma, but a palpable sense of the indelible marks war leaves on a soul. This initial conversation, the gentle unfolding of what lies beneath the surface, felt like a window into an entire world of sacrifice and resilience that most of us, me included, rarely truly acknowledge. It was a prelude to a storm delivered with the serene calm of its aftermath.

As he talked about the camaraderie and the crushing loss of brothers in arms. I envisioned the vast landscape of their collective experiences: the heat of distant lands, the ever-present danger that sharpens every sense to a painful edge, the moral complexities inherent in conflict, and the grief of losing those with whom you'd faced certain death.

I thought of the psychological wounds that are often far more subtle than the physical ones. The nightmares that persist long after the missions cease, and the struggle to reconcile the person they were before deployment with the person they became in its aftermath. This pain was a recognition of the profound cost of freedom borne not by nations or abstract ideologies, but by individual hearts and minds. It was the understanding that for many the war doesn't end when they step off the plane; it merely transforms into a new kind of battle fought on the home front with a battle against invisible enemies like PTSD, crippling moral injury, and the ghost of their past selves. It's the unforgiving burden of memory, the weight of responsibility for events permanently etched in a warrior's mind, and the quiet despair of feeling misunderstood by a world that never truly saw what they saw.

Alongside this ache, a thread of optimism emerged as vibrant and undeniable as the sunrise after the longest night. It wasn't a naive hope that war could be undone or that suffering could be erased; it was a confidence in the resilience honed by adversity into an unbreakable core. This expectation was present in Corbin's unwavering determination to share his story, not for pity, but for understanding to help build a vital bridge between his world and ours. It lay in the spirit that allowed Corbin and countless others to face unimaginable horrors and still find meaning, purpose, and a way to move forward, often in renewed service to others. This hope resided in the bonds of brotherhood and sisterhood formed in adversity and relationships that often transcend blood and endure long after service ends. These bonds are critical lifelines in the often-isolating civilian world. Through sharing these stories, it is a hope that a greater collective consciousness might awaken, leading to better support systems, deeper empathy, and genuine, lasting healing for our veterans. Corbin's ability to articulate, even hint at, such profound experiences while still maintaining a gentle composure was a beacon of that enduring human spirit, a testament to humanity's capacity to not merely survive, but to transcend.

The journey of military personnel returning to civilian life is complex and challenging. The transition is not a simple return to normalcy, but a pilgrimage marked by invisible wounds and difficulties in reintegrating into civilian life. Many struggle with feelings of isolation as they try to reintegrate into a society that cannot comprehend their experiences.

To truly support our troops, we must move beyond superficial gratitude and engage in sustained and meaningful support. We must advocate for robust and accessible mental health resources, facilitate meaningful employment, foster genuine community connections, and simply listening without judgment. We must recognize that healing is a journey, not a destination, and that our role is to walk alongside them, providing the resources and reassurance that they are not alone.

Gold Star families who have lost a loved one in active military service are forever changed by their sacrifice. The distinction of being a Gold Star family is one that no one seeks, it is a membership granted through profound loss. As the nation mourns the loss of a service member, the family's grief is both a public and private reality. The pain of losing a loved one is relentless, and the family's struggle to cope with their loss is often overlooked once the initial outpouring of sympathy subsides.

Corbin's narrative underscores the enduring pain and courage of Gold Star families. Their grief is not a temporary state but a permanent void that is a constant reminder of what was lost. We must commit to honoring their loved ones' legacies not just with ceremonies, but with ongoing community, remembrance, and support. This means acknowledging their pain, providing resources for bereavement and mental health, and ensuring that their fallen heroes are never forgotten. In honoring their sacrifices, we must recognize that the struggles are not just individual, but also collective. We, as a society, have a responsibility to ensure that their burdens are shared, their pain is acknowledged, and their love is remembered. By doing so, we can work towards creating a more supportive and inclusive environment for those who have served and those who have lost loved ones in service.

Corbin's story challenges us to reexamine our understanding of sacrifice, service, and support. By working together, we can create a more just and supportive society, one that honors the sacrifices of those who have served and those who have lost loved ones in service.

As we ended our ZOOM meeting the world outside felt a little sharper, a little heavier with the weight of Corbin's unspoken past but also infused with a fierce sense of purpose. His story isn't just his; it's a vital chapter in our history, a resonant echo of countless other sacrifices. It's a call to action, a gentle but firm invitation to open our hearts and minds to the profound costs of war and the incredible resilience of those who bear its weight, both on the battlefield and on the home front. It challenges us to listen more intently, support more actively, and remember more deeply, ensuring that the pain and hope embedded in every soldier's journey, and every Gold Star family's loss, never fade into silence.

I feel privileged to capture the wide range of emotions within this book and thank Corbin for his willingness to share his heartfelt story, for allowing us a glimpse into the landscape of his soul. Let the words of this book inspire you to support not only our military personnel, but also the Gold Star families wrapping them in a blanket of enduring remembrance and community. Their sacrifices, their resilience, and their continued struggles demand nothing less than our unwavering attention and our deepest empathy.

SSgt Corbin Sipe, United States Marine Corps

It is through listening to these stories, truly hearing them, that we begin to heal, not just the wounds of war, but the fractures in our collective human spirit.

Geoffrey Yoder – Author

Chapter Two

The Silent Oath

In a world where the digital noise and the relentless pursuit of validation have become the norm, it's easy to get caught up in the illusion that visibility is the ultimate measure of success. We're constantly bombarded with curated highlight reels and sensational headlines, each one screaming for our attention and begging us to hit the like and share buttons. It's a never-ending cycle of self-promotion, where the loudest voices and the flashiest displays of achievement are often mistaken for the most meaningful.

But beneath the glossy façade and distractions of social media, a profound truth lies waiting to be uncovered. It's a truth that reveals that the most transformative insights, the most resonant lessons, and the most lasting achievements are often found beyond the glitz of public spectacle and the fleeting trends of the moment. It's a truth that speaks to the power of quiet resilience, unsung commitment, and unwavering integrity.

Corbin is a man who embodied this truth. For years he had watched and observed as others achieved their own dreams and now it was his turn to take the leap. With a sense of purpose and determination, Corbin made the bold decision to join the United States Marines.

His decision to join the Marines was not driven by a desire for fame or fortune, but by a deep-seated sense of purpose and duty. He was drawn to the values of the Marine Corps, honor, courage, and commitment, and he was eager to be part of an organization that embodied these principles. Corbin was struck by the sense of camaraderie and a feeling of pride that existed among his fellow recruits. They were a diverse group of individuals but were united by a shared sense of purpose and a willingness to push themselves to the limit.

Corbin's story is a testament to the enduring power of quiet resilience and unsung commitment. It's a reminder that true achievement is not measured by likes and shares, or by public accolades and recognition. Instead, it's measured by the depth of our character, the strength of our convictions, and the positive impact we have on those around us. As we navigate the chaotic landscape of modern life, it's easy to get caught up in the noise and the distractions. Corbin's story encourages us to look beyond the surface level, to seek out the quiet power that lies within, and to cultivate the qualities of character that will truly make a lasting difference in the world.

This is the story of SSgt Corbin Sipe, United States Marine Corps

Corbin's quietness was a nuanced art, a deliberate withdrawal from the chaos that often defines adolescent interaction. It was not the silence of disinterest, but an active choice, a refusal to engage in the performative landscape of high school. While others vied for attention, Corbin cultivated the mastery of sustained observation. This reserve, compounded by the often-overlooked position of a middle child sandwiched between the more grandiose trajectories of his brothers, honed his skills in the subtle art of being overlooked.

To those who skimmed the surface, Corbin's silence could be easily misinterpreted. It might be mistaken for timidity, a detachment from the vibrant world or a lack of ambition or personality. Society often projects loud expectations, operating under the flawed assumption that desires unvoiced are desires non-existent.

Peering beyond the superficial gloss of his reserved demeanor revealed the profound error of such assumptions. Within this sanctuary resided his deepest thoughts, his most complex emotions, and his most ambitious aspirations, safeguarded from the clamor and casual judgment of the outside world. There, the true weight of his purpose was held, not a fleeting whim, but something profound and unwavering, akin to the deep, unseen currents that navigate a ship. This purpose was protected from the distractions of needing external validations. It was a power that needed no broadcast; it simply existed awaiting its moment to manifest with a quiet, undeniable force.

Understanding Corbin's drive, a force that has shaped his life, requires a deeper look into its roots as a childhood marked by a violent temper. Even at a tender preschool age, he struggled to control his impulses, often lashing out at those around him. His mother wrestled with how to steer him towards a more positive path, frequently turning to prayer for divine guidance.

A pivotal moment occurred during a Sunday school lesson that altered Corbin's trajectory. Witnessing a fellow student picking on another child, and seeing the indifference of those nearby, something within Corbin snapped. As his own temper surged, he instinctively intervened, seizing the aggressor's hand and forcing it into his own nose, causing it to bleed. In the ensuing quiet as he awaited the inevitable reprimand, tears welled up not from remorse, but from a bewilderment at his own actions. His mother was called to address his behavior.

A deeper conversation unfolded between mother and son. The discussion, rooted in love and a desire for change, ultimately led Corbin to embrace Christ as his savior. This spiritual awakening transformed his attitude, but the innate drive that had once manifested as uncontrolled rage remained. However, it was now tempered, channeled, and ultimately, controlled, setting the stage for the purposeful individual he would become.

The Grace of the Pause

For many families, the decision that Corbin would need to revisit his ninth-grade year would have felt like a crushing shame, a public marker of delay. It was the visual evidence of a track diverged.

But for Corbin, and for those who understood the intricate, non-standard wiring of his mind, the year wasn't a repetition; it was a renovation. It was anything but a distraction. Instead, it unfolded as a period of reflection, a fertile ground where fragmented cognitive knowledge, previously scattered on the surface, could truly take root and flourish.

We are taught to imagine high school as a rigid, linear sprint, a desperate race against the clock to accumulate credits and advance. Corbin, facing the reality of the redo, was suddenly given the unexpected grace of a pause. The initial sting of the decision, perhaps a whisper of frustration or the generalized ache of feeling "behind," quickly dissipated. Without the bustling, often overwhelming pressure or the exhausting ritual of constant peer comparison, Corbin found something invaluable he hadn't had before: space. Space to breathe, to think deeply, and, most importantly, to truly learn.

His mother, sensitive not just to the academic necessity but to the emotional weight of revisiting material, encouraged him not to simply repeat the year, but to actively re-explore it.

What had once felt like a frantic scramble to memorize facts and formulas transformed into an intellectual excavation. The learning environment shifted from a performance stage to a quiet sanctuary. History, which had been a parade of names and dates to be transcribed, became a living, breathing thread of human intention and consequence. He delved into the "why" behind events, tracing cause and effect that the traditional classroom's syllabus rarely afforded. He wasn't just learning what happened; he was understanding the forces that drove humanity forward, granting him empathy for the past.

This second pass at ninth grade became an intimate dialogue with knowledge itself. When a concept didn't click immediately, there was no paralyzing rush to move on to the next chapter simply because the calendar demanded it. Instead, he could sit with it, wrestle with it, approach it from different angles, research it online, or discuss the nuances at length with his parents. This wasn't remedial work; it was profoundly foundational. He was learning how he learned best, discovering his unique intellectual landscape, mapping the pathways where understanding naturally flowed.

What emerged from this quiet year was not just better grades, but a quiet, unshakeable confidence, a deep-seated understanding that went far beyond recitation or recall. His cognitive knowledge wasn't just growing in quantity but in quality, deepening its roots with every new connection made. He was developing genuine critical thinking skills, a natural inclination to question, to synthesize, and to build intricate mental models rather than holding fragmented pieces of information. The perceived "distraction" of redoing a year became a vital period of integration, allowing disparate facts to coalesce into a cohesive, meaningful worldview.

Looking at Corbin now, one wouldn't see a boy who was "behind." Instead, you see a young man with a thoughtful, measured gaze, an ability to articulate complex ideas with profound clarity, and a genuine, lasting love for learning that had been nurtured in the quiet sanctuary of his reflective year.

His journey serves as a heartfelt testament to the idea that sometimes, the most profound steps forward are taken when we grant ourselves the time to pause, to truly understand, and to allow growth to unfold at its own, beautiful, and necessary pace. That outwardly backward step was, in fact, the launchpad for a much richer, deeper, and ultimately more meaningful educational experience.

The Calling: The Few and the Proud

Corbin's path to the United States Marine Corps was not a sudden epiphany, but rather a combination of different experiences, introspections, and encounters that, over time, converged into an undeniable calling. His enlistment with the United States Marine Corps was the culmination of these different events, each one in the narrative of a young man discovering his true potential and his destined purpose.

He harbored a burning ambition to prove he was not only tough enough to succeed where others might falter but also capable of achieving more than what he perceived others expected of him. This yearning for recognition, for an arena where his strength and determination could be displayed, became the bedrock of his quest for significant challenges. He wasn't just looking for an accomplishment; he was searching for a transformation, something in which his character could be forged anew.

The first tangible spark arrived during his junior year of high school amidst a deluge of recruitment brochures from various military branches. While many of his peers might have casually discarded them, he approached these pamphlets with a nascent curiosity, a subconscious search for the very challenge he craved. It was a Marine Corps pamphlet that truly captured his attention. Inside, a series of pointed questions challenged the reader: "Do you have what it takes to be a Marine?" He, reviewing each query, found himself answering yes to more than the requisite five. It wasn't just the sheer number of affirmative responses that resonated; it was a bold, declarative statement within the brochure: "Do you have what it takes to be a Marine?" This wasn't merely a question; it was a direct challenge to his core motivation, a gauntlet thrown before his desire to prove he was better, tougher, more capable. The pamphlet articulated the very essence of the proving ground he sought, locking the image and the question firmly in his memory, a silent promise of future exploration, even if the decision to sign up remained distant.

The next significant moving event occurred when he was watching President George W. Bush's final State of the Union address. Amidst policy discussions and reflections on his presidency, President Bush paused to acknowledge an extraordinary act of heroism: a Marine who had risked his life to save fellow Marines. As the camera zoomed in on the lauded Marine, Corbin was transfixed. It wasn't just the bravery that struck him, but the man himself. He couldn't help but notice the Marine's demeanor, a quiet strength, an undeniable gravitas, a settled confidence that spoke volumes without a single word. There was an aura of purpose, of disciplined self-possession, that deeply impressed Corbin. In that moment, a powerful, aspirational thought solidified in his mind: "I want to be like him." This image of quiet heroism, embodying the very qualities of resilience and sacrifice he admired, further cemented the Marine Corps as a potential path. Still, despite this profound impression, the commitment to enlist remained an unmade choice, waiting for further confirmation.

The journey continued through an unexpected detour: a medical emergency during his senior year of high school. Corbin was diagnosed with appendicitis, necessitating surgery for its removal. While recuperating in the hospital, he received a visit from the hospital chaplain. During their conversation, Corbin, perhaps testing the waters or simply sharing a casual interest, briefly mentioned his curiosity about joining the Marines. To his surprise, the chaplain revealed that he, too, had once served as a Marine. This revelation left an impression on Corbin. The chaplain, far from fitting any stereotypical image of a hardened warrior, possessed a quiet, easy-going demeanor, a gentle strength and profound compassion. This unexpected blend of military background and serene temperament further deepened Corbin's respect for the Corps, demonstrating that the term "Marine" encompassed a broader spectrum of character than he had initially imagined. The chaplain's calm affirmation, "You would not regret joining the Marines if that was your decision," served as another powerful endorsement, a personal testimony that resonated deeply with Corbin's evolving considerations. Even with this intimate and reassuring encounter, the decision to enlist remained elusive, a weighty commitment still pending.

The recurring theme of Marine influence continued to weave its way into Corbin's life. Not long after, while riding with a co-worker to a job site that required overnight stays, the conversation turned towards military service, specifically the reserves. During their interactions, the co-worker casually revealed his own background: he was a former Marine who had later transitioned into the reserves. This was not a recruitment pitch, but a simple, almost serendipitous disclosure that reinforced the consistent presence of Marines in Corbin's orbit. It was as if the hand of providence itself was subtly directing him, placing individuals who embodied the Marine spirit in his path. "Seemed like one event after another," Corbin reflected, "kept pointing to a future with the Marines." Each encounter, each casual conversation, each profound observation, progressively tightened the threads of his conviction.

Beyond these pivotal personal encounters, the direct influence of a Marine recruiter also played a persistent, practical role. The recruiter continued to engage with Corbin, providing information, answering questions, and even visiting his home to speak with his mother about his prospects. This consistent presence, combined with the internal reflections and external confirmations, allowed the idea of joining the Marines to remain at the forefront of his thoughts. The cumulative weight of these experiences; his initial desire for challenge, the pamphlet, the inspiring Marine on television, the serene chaplain, and the unassuming co-worker, had built to an undeniable crescendo.

Crossing the Decisive Line

In February 2010, the combination of events, feelings, and influences coalesced into a firm resolution. Corbin enlisted in the United States Marine Corps. His desire to prove he was "tough enough to succeed where others could not," to find an environment where he could "accomplish more than he felt others expected," found its ultimate expression in the demanding world of the Marines. For Corbin, joining the Corps was more than just signing up for military service; it was an active commitment to a journey of self-discovery, a deliberate step towards forging an identity defined by resilience, dedication, and the unwavering pursuit of excellence, finally embracing the challenge he had been seeking all along.

The air around him seemed to thicken, charged not merely with humidity but with the profound, static anticipation of radical change. As Corbin stepped forward, his boot sole meeting the unforgiving concrete, he crossed a silent, decisive line, a chasm separating a familiar, quiet past defined by personal autonomy. In that singular instant, the casual rhythm of his civilian life was irrevocably changed to the relentless, structured intensity and controlled chaos of the Recruit Depot. He paused only long enough to allow the profound magnitude of his choice to settle over him, an invisible but heavy weight, prefiguring the unfamiliar fabric of the uniform he would soon wear.

This was no conventional occupational transition or typical career trajectory. The act of crossing the threshold was, for Corbin, a sacred, binding ceremony, a vow of allegiance to an entity far larger than himself, greater than the sum of its individual parts. He was not simply embarking on a difficult profession; he was stepping, boots-first, into a formidable, sacred legacy defined by the supreme historical imperative of sacrifice, constructed on the unshakeable bedrock of unwavering duty, and sustained by an enduring, fiercely maintained spirit of selfless service. He was pledging his very being, his physical capacity, his mental acuity, and his moral compass, to the defense of abstract principles forged in blood, preserved through relentless institutional discipline, and held aloft by the collective will of dedicated individuals. The uniform he was about to don represented centuries of such commitments, a mantle of honor demanding absolute fealty.

The weight of this inherited legacy, the expectation to uphold faith, not just in country, but in the Marine standing to one's left and right, is what truly defines the ordeal. It is a weight that immediately begins the process of stripping away the superficialities of ego and personal vanity, demanding a rebirth that prioritizes the mission above all else. This process, brutal and necessary, often cracks those who mistakenly believe physical prowess or loud confidence is sufficient. But Corbin, possessing a deep understanding of himself, knew his true foundation lay elsewhere.

His quiet nature, far from being a perceived liability in a world that often rewards volume, bravado, and self-promotion, was, in fact, his inherent and meticulously honed strength. He possessed a vital ability to observe and process, to listen not just to the clipped, immediate words of command, but to the often unspoken, higher intent behind them. His mind was unclouded by the internal static of personal ambition or defensive pride. Unlike those who fight the system with rhetorical resistance or emotional defiance, Corbin was already adept at submitting completely to the codified language of command, following orders with immediate, precise action. He prioritized the monumental significance of the mission above any personal comfort, fleeting vanity, or physical strain.

He knew that the true art of military service demands a disciplined mind that can filter the noise of high-stress environments. In the blur of combat or crisis, the individual who can quiet the instinctive panic, who can recall the training sequence buried deep beneath the surface of immediate fear, is the one who survives—and more importantly, the one who ensures the survival of the team. This ability to maintain internal equilibrium under extreme pressure is the direct byproduct of the focused silence Corbin had always embraced. His strength lay, paradoxically, in his very humility—the willingness to learn, to fail privately, and to rise again without the need for public validation.

The oath he had taken was a societal transaction of the most serious kind. It was a covenant that placed the needs of the many squarely before the wants of the one. To Corbin, and to all the quiet, dedicated young individuals who bravely choose the hardest paths—who voluntarily leave the comfort of the known world for the rigorous, often thankless, defense of an abstract ideal, who understand intrinsically that

true, lasting freedom is not a spontaneous birthright casually received, but a hard-won achievement, earned daily through the exacting, continuous tolls of discipline, perseverance, and sacrifice

Chapter Three

Boot Camp

For those who choose the demanding path of a Marine, the journey begins not on foreign soil, but in a focused environment designed to test the absolute limits of human endurance: the test of commitment. The sterile, impersonal nature of barracks and temporary housing replaced the warmth and familiarity of a childhood bedroom, and the concept of "home" transformed from a physical place into a collective memory, a shared hope, and the enduring connection with those who understood the unique pressures and demands of the Corps. The intricate dance between the warrior's duty abroad and the civilian's life back home created a constant required immense strength to navigate. The leather of the boots ever touched the soil of a foreign shore, before the structure of strategic missions unfolded across continents, and long before the relentless rhythm of operational tempo became the defining pulse of their existence, there lay the trails of transformation: training.

It was within this intense service that the fundamental principles of discipline are not just taught but etched into the very souls of recruits. A precise and potent lexicon emerged, a language of unwavering focus, of constant, ingrained readiness, and of immediate, unquestioning execution. This environment, though undeniably harrowing, both physically and mentally, served a higher, more profound purpose. Beyond the sweat and the strain, it was the fertile ground where the potent, unbreakable bonds of brotherhood were organically forged. This was the genesis of an unshakeable loyalty, a shared understanding born from the collective experience of facing uncertainty with bated breath, of pushing beyond the perceived limits of exhaustion, and of learning from the sting of failure together, only to rise from the ashes, stronger, more unified, and inextricably linked by the shared ordeal.

One aspect of Corbin's training after Boot Camp that served him well in Afghanistan was his Infantry Training. Infantry training is far more than a checklist of physical drills; it is a comprehensive psychological and physiological program that forges stamina, skill, mental resilience, and an unbreakable bond of teamwork into one highly functional organism. The crack of rifle-fire across a battlefield presents a picture of disciplined order. In the chaos of conflict, the sight of infantrymen moving in unison, a synchronized line of advancing Marines, a team flawlessly providing cover fire, commands a sense of awe. This polished performance, however, is merely the visible façade of an immense and deeply human iceberg. Beneath the surface lies months, sometimes years, of systematic and grueling training designed to transform individuals from different civilian backgrounds into a single, cohesive, lethal, and adaptable fighting force.

This exhaustive, demanding regimen was the essential, non-negotiable groundwork upon which true resilience was built. It meticulously prepares the physical vessel to endure profound hardship and honed the mental acuity to navigate the treacherous currents of complex, high-stakes scenarios where lives hang in the balance. Even with its formidable intensity, the physical preparation, while crucial, could only carry him so far. It was fundamentally incapable of inoculating the heart against the state of perpetual readiness that became his constant companion, nor could it fully prepare him for the ceaseless, demanding motion that defined a life of global deployment, and certainly not for the profound emotional distance that, like a shadow, became an inescapable aspect of his existence. The internal shifts, the quiet erosion of personal boundaries, and the gradual recalibration of what "normal" felt like were the invisible casualties of this rigorous path.

The weight of global responsibility, though often unspoken, was a constant presence. Corbin, like every Marine, understood that his actions, however small, were part of a larger international relations and security activity. This understanding filled his service with a profound sense of purpose, far beyond personal ambition. It was the knowledge that he was a cog in a vast, complex machine designed to protect interests, uphold values, and provide a bulwark against instability. This elevated sense of mission, while a powerful motivator, also served to further detach him from the everyday concerns that preoccupied those outside the military sphere. The mundane anxieties of civilian life often seemed trivial when compared with the stark realities he faced.

The concept of sacrifice permeates every aspect of a Marine's life. It was not limited to the dramatic act of laying down one's life, but extended to the daily relinquishing of personal desires, the deferred dreams, and the missed milestones. Birthdays passed without celebration, anniversaries are observed from afar, and the simple comfort of routine is a luxury rarely afforded. This state of selfless devotion also chips away at Corbin, carving out space for Corbin felt alien to the person who had initially chosen this path. The intimate relationships bore the brunt of this constant strain, necessitating an extraordinary level of understanding and fortitude from loved ones.

Chapter Four

Deployment

For us, as civilians, the word "deployment" often glides off the tongue with a clinical, almost abstract sterility – a temporary absence, a necessary duty. But for a Marine like Corbin, a deployment is anything but sterile; it is a radical, almost violent disruption of life as he knows it. It's not a mere pause, but an absolute, if temporary, severing of established ties - the comfort of routine, the embrace of family, the predictability of home. During these extended moments, life shrinks to the immediate, hyper-focused world of his platoon, squad, or fire team – where every breath, every decision, rests on the collective; yet simultaneously, it expands to encompass the overwhelming complexity of international politics, unfamiliar geographies, and the high-stakes consequences of global events. To truly honor this dedication, we must look beyond the mission success reports and analytical summaries and focus empathetically on the human element. The true weight of a career defined by movement is compounded by the cost of distance.

The Marine is trained to focus intensely on the immediate mission at hand; that fierce dedication is non-negotiable for success and survival. But the heart, even when compartmentalized, never fully disconnects from the life left behind. There exists a deep, constant, and often unspoken tension between the absolute commitment to duty and the profound, aching yearning for home.

This constant oscillation between two radically different realities, the high-stakes vigilance of the deployed environment and the peaceful, continuing normalcy of the life left behind, is an immense emotion burden.

The emotional distance that characterizes a life of global deployment is perhaps the most difficult to bear. These global movements meant calculating time differences just to catch a fleeting, often delayed conversation with loved ones before their bedtime. It meant missing the quiet but precious milestones: the birthdays, the anniversaries, the first steps, and the mundane but essential moments of daily life that quietly accumulate into years.

Corbin became a master of compartmentalization, focusing entirely on the task at hand to ensure collective safety. The emotional processing is merely delayed; it is never erased. The weight accumulates, creating an inner landscape defined by both fierce capability and quiet, enduring sacrifice. The inherent nature of military service required Corbin to compartmentalize his emotions, to maintain a stoic façade in the face of danger and adversity This learned resilience, while essential for survival and operational effectiveness, often creates a barrier to genuine emotional connection back home. The shared experiences of combat, the camaraderie forged in the fire of battle, are experiences that civilians could never fully comprehend. This creates an isolating effect, a sense of being perpetually out of sync with the world he was sworn to protect. His experience as a Marine quickly unspooled into a complex, multi-layered intricate design. This wasn't merely a geographical journey across disparate time zones and contrasting continents; it was a profound psychological and existential navigation through vastly different realities, each demanding rapid and total recalibration of self. These burdens are not singular; they are intensely tactical, demanding peak physical and mental acuity in lethal environments; they are deeply cultural, requiring rapid comprehension and respect for unfamiliar norms; and they are profoundly spiritual, forcing introspection on purpose, sacrifice, and the raw realities of conflict and human suffering.

Corbin, like every Marine, had to master not only the immediate, often life-or-death objectives of the mission at hand, but also simultaneously maintain a keen, almost instinctive awareness of the broader world's intricate geopolitical stakes. This mental tightrope act unfolded constantly, thousands of miles from the comforting, familiar anchors of home-a psychological distance that often felt far greater than any physical measurement.

We did not talk about every aspect of Corbin's deployment during our ZOOM and telecon tag-ups. His boots may have crunched on the shifting sands of a desert landscape, where the blurring sun bleached all color from the world, and the only visible horizon was the often agonizingly fine line between un-wavering duty and profound exhaustion. For one memorable deployment, the arid dust of Afghanistan clung to everything - uniforms, equipment, even the phantom sensation of it coating the back of one's throat, tasting of ancient earth and the unyielding sun. The desert, with its vastness, often concealed unseen threats, demanding an almost telepathic awareness of one's surroundings. For Corbin, his deployment to Bagram Air Force Base was a mission that carried a weight far beyond the grit of the landscape. It was of duty, brotherhood, and the quiet, persistent effort to reshape America's role in a land steeped in complexity. More than just a tour of duty, it was where profound bonds were forged, where the measured rhythm of training alongside the Georgian military intertwined with the harsh drumbeat of war's enduring realities.

He was on a 7-month deployment to Afghanistan as part of a seven-man Marine liaison team working with the [1] Georgian military to train with the Georgian forces participating in missions with them as additional security forces to try to reduce the US footprint in Afghanistan. This was a unique

---

1. The Georgia Defense Readiness Program – Training (GDRP-T) was a bilateral United States-Georgian training program launched on May 1, 2018. Credit: Defense Media Activity – WEB.mil

unit tasked with an advisory role—a strategic divergence from the frontline combat missions that typically defined Marine Corps deployments. They were the subtle architects of transition, the steadying hands guiding a burgeoning partnership, always acutely aware of the volatile dangers that simmered just beneath the surface. Their presence wasn't intended for battlefield domination; it was about empowering allies, fostering self-sufficiency, and meticulously dismantling the entrenched layers of a long-standing intervention. Initially deployed in July 2018, their team was augmented a month later, filling newly designated roles, each man a crucial cog in this small, vital mechanism. Together with their Georgian interpreter, they formed a tight-knit core, unified by their shared objective of seamless collaboration and responsible drawdown. Their days unfolded as a complex mosaic of intricate logistics, nuanced cultural exchange, strategic tactical planning, and the ceaseless vigilance demanded by their inherently hazardous environment. Every handshake, every lesson imparted, every joint patrol was a deliberate step in a larger, delicate dance of withdrawal and enduring partnership.

Team 4, Call Sign Creeper

Within the demanding environment, a profound camaraderie flourished. It transcended the predictable bonds that naturally develop among Marines; it was an accelerated intimacy, born of constant proximity, shared minor discomforts, and the pervasive, unspoken understanding of ever-present danger. Corbin's candid words paint a vivid picture of this unwavering mutual support: "Teamwork, rely on each other. Everyone has each other's back. We had a mission to fulfill and are all trained to accomplish the mission." This sentiment went beyond mere professional respect; it was the deep, unspoken understanding that arises from confronting shared vulnerabilities, a family forged not by blood ties, but by the relentless pressures of a combat zone. They shared meals, found solace in stale jokes, grumbled about the oppressive heat, debated intricate strategies, and instinctively picked up the slack for one another without a word needing to be spoken. "While honoring command structure, we treated each other as family, relying on each other's skills. We got along well, respected each other." The formal hierarchy of the Marines softened into a more organic, familial structure within their small team, each man placing implicit trust in the competence and unwavering dedication of the others. The interpreter, a vital bridge between cultures and languages, became an indispensable member, his keen insights and timely humor often serving to diffuse tension, further cementing their unit's unique and cohesive identity.

The Contingency of Loyalty

This familial bond was a daily test, challenged not only by external threats but by the stark, internal contemplation of what "family" truly signified in such a precarious setting. They confronted the inherent, brutal risks which would likely be disarming to anyone outside their immediate experience. Their training extended far beyond conventional combat scenarios, delving into the agonizing possibilities of injury or loss within their close-knit unit. It wasn't sufficient to be merely prepared for the enemy; they had to be prepared for the unthinkable permutations of their own ranks being compromised.

"Needed to train for an event if one of the team was injured and could not fulfill their task. Needed to train for succession," Corbin explained, a somber note in his voice that hinted at the profound weight of these exercises. "This was a humbling experience needing to work through these scenarios. Brothers in arms. Sobering to work through the possibility of injury or death to continue the mission." They would convene with maps spread out, their faces etched with grim determination, discussing the unspeakable contingencies. These weren't theoretical skirmishes confined to a training manual; these were detailed, meticulously planned protocols for the actual, terrifying moments they knew could materialize. They painstakingly planned for every horrifying contingency: who would seamlessly assume the turret position if the designated gunner fell, how the mission would continue if a critical member was irrevocably lost. They practiced the grim choreography of casualty evacuation, the immediate, almost instinctive shift of responsibilities, the seamless integration of backup roles even as the prospect of grief and shock threatened to paralyze.

This wasn't a cold, detached exercise in strategic detachment; it was an act of profound loyalty, unspoken agreement to persevere, to honor the fallen by seeing the shared purpose through to its conclusion. They faced the unimaginable, not with crippling fear, but with a quiet, resolute determination to protect each other and their overarching mission. Each man understood implicitly that if he were to fall, his brothers would not falter; they would readily pick up his fallen duties and carry the mission forward to its successful completion. This shared understanding was the very bedrock of their collective resilience, a grim but essential promise whispered in the dust-laden air of Afghanistan.

Life is woven with countless threads of choice and circumstance, but sometimes, a single, unseen hand tugs at a thread, altering the entire pattern forever. We call it fate, a mysterious force that operates beyond our comprehension, often revealing itself not in grand pronouncements, but in the softest, most unexpected whispers. In the relentless ordeal of war, where lives are measured in fractions of seconds and decisions carry monumental weight, fate's whispers can be deafening in their implications, carving a canyon between existence and oblivion.

For Corbin and his brothers in arms, their deployment had become a grueling rhythm of patrols, and the ever-present hum of danger. Week by week, the countdown to their return ticked down, each sunrise bringing them closer to home, to the familiar embrace of loved ones. The camaraderie forged in shared hardship was a palpable shield against the harsh realities outside the wire, a bond tighter than blood. They were a family, dependent on each other, moving as one.

Then, just weeks before their deployment was to end, tragedy struck. April 8, 2019, dawned like any other day, the sun rising relentlessly over the arid plains, promising another sweltering cycle of duty. But for Corbin, it held an unseen hand, a subtle shift in the meticulously planned schedule that would redraw his entire destiny. As the primary driver, his presence on missions was almost a given, a constant, reliable fixture. Being in the driver's seat meant he was vital to every movement, every patrol outside the wire, his hands on the wheel, his eyes scanning for threats, his instincts honed by countless hours of relentless vigilance. That morning, scheduled to attend a training course whose duration was uncertain, he was left off the mission card.

It was a seemingly insignificant administrative decision. A momentary, mundane detail in a world where lives hung in the balance of fractions of a second. He might have felt a flicker of annoyance, a slight relief to step away from the daily grind of driving, or perhaps nothing at all, merely accepting the shift as a necessary part of the military machine. He had no inkling that this fleeting scheduling quirk, this bureaucratic anomaly, would serve as his improbable shield. But it was a decision that would ultimately determine his fate and the tragic reality that awaited his brothers. His absence, determined by a fleeting schedule, would, in an instant, become the thin, fragile line between his life and unimaginable loss.

He was in the temporary safety of the base, tucked away from the unpredictable violence of the mission field. He was likely focused on the task at hand, his thoughts far from his team, perhaps even looking forward to catching up with them later, to swapping stories from the day's events. But for his teammates—his brothers—the routine mission turned instantly catastrophic. Their vehicle with Corbin's teammates was struck by another vehicle carrying an explosive equivalent to a crippling 600-pound car bomb. The immediate, brutal force of the blast ripped through flesh and steel, claiming the lives of the driver, assistant driver, and turret gunner. Three of his brothers, gone in a deafening, blinding flash. The very roles he would have occupied, extinguished in an instant.

Chapter Five

The Agonizing Wait

The news arrived in a slow, disorienting haze. Corbin was asleep after completing his required training, a brief reprieve from the relentless schedule, unaware that his world had already begun to shatter miles away. He woke to the unusual lack of the usual sounds. The air even felt different, heavy with an unspoken truth.

"As I walked to the chow hall, everything seemed quieter than before. Normalcy was off; it felt weird that something must be wrong. Vehicles were gone, which was unusual," he recounted. The distant hum of generators, typically a constant thrum, seemed muted, stifled. The usual morning chatter, the clang of equipment, the distant shouts of training, were all strangely absent, replaced by an unsettling hush that stole the oxygen from the air. The familiar line of his team's vehicles, usually parked in their designated spots, was hauntingly empty. An icy dread began to coil in his gut, a premonition that something terrible had happened, a feeling that defied logic but resonated with the instinctual alarm bell of a soldier in a combat zone. He knew, with a certainty that something was profoundly wrong.

The sense of dread sharpened into an agonizing certainty when he was told the devastating truth: a vehicle had been hit, and there were fatalities. Time seemed to stop. The words hung in the air, heavy and poisonous, each syllable a hammer blow to his soul. In that immediate moment of shock, before he could process the names or the severity, before the full horror could settle, Corbin's instinct was pure and primal. He grabbed his phone and sent a quick message to his girlfriend, now his wife: "Love you. I can't talk now. We're getting a call. I gotta get going. I love you like crazy honey. Love you." It was an instinctive tether to life, a desperate anchor to everything good and whole, before rushing headlong into the uncertainty of death, into the grim validation of every contingency plan.

For hours, he endured the cruelest form of military agony—the wait. He teamed up with another unit, rifle in hand, driving toward the base entry point. The rifle in his hand was less a weapon and more a physical manifestation of his profound helplessness and readiness to engage whatever threat had taken his team. He was ready to fight but also vengeance was the last thing on his mind. He had equal parts of fear that because his brothers were just killed, the same might happen to him. The only battle left was the internal one, grappling with the terrible truth that the meticulously planned succession training had become a necessity, and that the family forged in the dust had been irreparably broken. The silence that had initially signaled something was wrong now screamed the unbearable cost of war, reverberating with the final, tragic whisper of fate that had spared him, but stolen his brothers. He stood there, a survivor burdened by an incomprehensible twist of fortune, his heart echoing the empty spaces left by the men who would never return.

Eventually, the heavy machinery began to throb closer, a sound that was meant to be a balm but instead sent a fresh wave of dread through Corbin. It was wrong. The rhythm was hesitant, burdened, lacking the confident, unwavering surge of vehicles carrying men who had faced the enemy and were now returning to the safety of the base. There was a subtle, terrifying discordance in the approaching vibrations, a note of sorrow woven into the mechanical roar.

Then came the grim confirmation. Corbin heard the new about Ben, Chris, and Rob being killed from another Marine in the unit as he was waiting by the staging area of the ECP (Entry Control Point) before the vehicles from the QRF (Quick Reaction Force) returned to base.

The truth hit Corbin with the stunning, undeniable force of a blast wave. It was a silent detonation within his mind, a catastrophic implosion of everything he held dear. His teammates, the men whose lives were inextricably interwoven with his own through shared hardship, stale jokes whispered in the dark, and an unwavering trust that surpassed blood ties— his chosen brothers— were gone. The loss was absolute, a void so profound it felt physical. It was overwhelming, and in its sheer magnitude, utterly isolating. They were not just casualties; they were holes torn in the fabric of his existence.

Remember SSGT Slutman, SSGT Hines, and SGT Hendriks

For one searing moment of pure, unadulterated humanity, Corbin turned away from the observation post, his gaze dropping to the scorched earth beneath his boots. In that instant, the rigid discipline of a Marine, the ingrained sense of duty, dissolved like sand in a flood. After hearing the news of their passing, he walked away from everyone and knelt down sobbing for what seemed like a long time but was probably only 30 seconds. Tears, hot and stinging, spilled from his eyes, and in the arid desert air, they instantly turned the fine dust on his cheeks into streaks of mud, an outward manifestation of the inner turmoil. It was a moment of raw, vulnerable sorrow, a testament to the depth of his love for the men he had lost.

But in the unforgiving environment of an operating base, such moments of profound emotional release were a dangerous luxury. He had a duty to fulfill, a mission that continued regardless of the personal cost. The danger that had just consumed his friends still lurked, invisible and menacing, just beyond the camp perimeter. The war did not pause for grief. It demanded vigilance, it demanded action, and it demanded that he, like the others, swallow the pain and stand ready.

The Crushing Silence of the Barracks

The emotional truce Corbin imposed upon himself held steady through the grueling remainder of the deployment. He functioned with a precision that bordered on robotic, a testament to his years of training and his ingrained sense of responsibility. Beneath the veneer of flawless execution, the truce was incredibly fragile, strained to its breaking point by the pervasive silence that now permeated the barracks. The laughter, the boisterous arguments, the familiar voices sharing stories and anxieties, all were conspicuously absent, leaving behind a hollow echo.

The empty bunks, the unused lockers, the spaces where camaraderie once thrived, these physical manifestations of absence intensified the atmosphere. The once vibrant, lived-in quarters were transformed into hollow monuments to the men who would never return. Every mundane task, every shared meal was a stark, unavoidable reminder of the immense weight locked away within Corbin's chest. He was a ghost in his own life, surrounded by the spectral presence of what had been lost.

The days that followed were a blur of grief, duty, and the haunting echo of "what if." Survivor's guilt, an insidious poison, began to seep into the quiet corners of his mind. Why him? Why was he spared when his brothers, men he had laughed with, fought alongside, depended on with his very life, were gone? The administrative course, once a minor detail, now loomed as the most significant turning point of his existence, a cruel twist of a bureaucratic thread that had severed him from their shared destiny. The whisper of fate had not been gentle; it had been a violent, decisive intervention, leaving him to grapple with the profound weight of an inexplicable grace.

Corbin continued to execute his duties, refusing to delve into the unspoken grief that festered amongst the survivors. He allowed himself to feel only the cold moving through the final weeks of the deployment in a trance of professional competence. He was a machine trained for war, and in the face of such profound loss, he retreated into the only operational mode he could access. He refused to look too closely at the contents of the emotional vault he had so carefully constructed; the fear of what lay inside was too great.

But beneath the surface, the containment structure was cracking under the immense, subterranean pressure of pain. The constant strain, the suppressed sorrow, the unspoken guilt of survival – it all pressed down, threatening to shatter the meticulously built defenses. The true reckoning, the emotional dam break, was merely postponed, a catastrophic event waiting for the right moment, the smallest tremor, to unleash its devastating flood. The silence, once just a noise, had become a suffocating shroud, a constant reminder that the war within him was far from over. The quiet hum of waiting had evolved into a deafening roar of unshed tears and unspoken goodbyes, a testament to the enduring weight of what had been lost.

Even now, years later, the memory of that day remains vivid, a silent testament to the arbitrary nature of life and death in combat. He carries not only the scars of war but also the indelible imprint of that whisper, a constant reminder of the men he lost and the life he was given. He lives with a heightened sense of purpose, a profound appreciation for every sunrise, and a determination to honor the sacrifice of his brothers by living a life worthy of their memory. The quiet on the base, the empty vehicle slots, the "I love you" message sent in desperation – these are not just memories, but an enduring whisper, a persistent question, and ultimately, a profound testament to the fragile line between fate and human existence.

Chapter Six

Unbreakable Covenant

The Breaking of the Dam

The full weight of the sacrifice did not settle upon Corbin until he was safely out of the operational environment, back in the relative peace of the United States. Stepping onto American soil brought a relief that was immediate and visceral: the primal, involuntary rush of endorphins signaling to the deep brain that the threat matrix had been dissolved. The survival response signaled the body was safe—the shoulders dropping, the hyper-vigilance momentarily dissolving, but that profound, clean relief was instantly soured. The clean, un-scorched air tasted of ash and guilt. Corbin's returning home was not to a hero's welcome, but to a void. The grief was a shard of hot metal lodged deep in his soul, and the well-meaning platitudes, "Time heals all wounds." You must move on" felt like a betrayal. How could he move on when part of him was anchored there, amidst the men who hadn't come back?

Stripped of the immediate, existential requirement of combat and survival, the carefully constructed walls around his grief began to crumble. These walls were built not just of emotional suppression, but of operational necessity: the need for focus, decisive action, and emotional compartmentalization. Now, the environmental pressure that demanded suppression was gone, and the sorrow burst forth with the force of a tectonic shift, a psychological shockwave that shattered his fragile equilibrium.

The homecoming was not a celebration; it was a devastating confrontation with the ghosts he had carried thousands of miles. The immediate emotional deluge was overwhelming: raw pain, confusion, and despair that manifested in sleepless nights and paralyzing stretches of silence. But intertwined with the pure, honest sadness lay a more corrosive wound: the haunting, relentless interrogation of the survivor.

"I should have been there with my team. I was the designated lead. Why was I spared from death? Why am I here, breathing this clean air, tasting this normalcy, when the best of them—the ones who deserved it more—are not?"

This profound survivor's guilt became inextricably linked with white-hot anger, anger that his colleagues were taken, anger at the senseless randomness of war, and a deep-seated personal responsibility for the vacuum they left behind. The transition from active duty to civilian life became less about decompression and more about navigating a profound, moral, and emotional crisis. He lived in a state of perpetual debt.

He found himself wrestling with the unbearable lightness of being spared, his mind obsessively replaying every decision, every movement of the days leading up to the incident. He conducted an agonizing, internal court-martial, searching for the fatal error he somehow missed, the fraction of a second where the universe had arbitrarily spared him while condemning his brothers. The vault of suppressed trauma had finally burst open, releasing not just grief, but a debilitating sense of injustice directed squarely inward, convinced he was a fraudulent survivor.

This torturous introspection, however, served as the beginning of a new, vital mission. From the depths of his anguishing guilt and pain, Corbin forged an unwavering mantra that became his lifeline: "We need to take care of their families. They are our ongoing mission, our continuing legacy."

This commitment became his anchor, the purpose born out of profound brotherhood and enduring loyalty. He could not bring his teammates back, but he could ensure their sacrifice was never forgotten and that their loved ones were continuously and tangibly supported. The survivor's burden transformed from passive, destructive pain into active, necessary commitment.

It was a mission to bridge the gap between those who served and those who were left behind, an attempt to honor the lost through constant, enduring action. He realized that the fierce loyalty forged in combat did not end when the caskets were lowered.

The mission to protect those families is, for Corbin, the continuation of the patrol, carried out on the home front of an enduring commitment to life, loyalty, and legacy.

Chapter Seven

Invisible Burdens of the Heart

The human heart, an intricate chamber of both boundless love and unbearable sorrow, carries burdens that are often invisible to the eye. For some, these weights are etched in the physical scars of battle but for others, the burden is manifested in other ways: the aching architecture of absence, a profound testament to the truth that trauma is not always measured by what the eyes witness, but by the silent, aching void the heart chooses to carry.

Corbin's journey into this unique landscape of grief began not with a direct impact but with a moment of shock delivered by a fellow Marine. Though physically removed the day his unit suffered devastating casualties in Afghanistan, with the seismic wave of that loss struck him with a persistent and heavy current of survivor's guilt. "As a result, I didn't experience the loss firsthand in that I saw anything," Corbin explains, reflecting on that fateful day. "I honestly think that's the Lord's hand in that. Nonetheless, I struggled and must process survivor's guilt." This sentiment, a quiet acknowledgment of both grace and a lingering wound, speaks volumes. It's the cruel, relentless "what-if" that haunts the quiet hours, an echo of the lives that ended too soon, and the life that, by some twist of fate, endured. The proximity of his own safety to their ultimate sacrifice became a torment, a constant whisper asking, "Why them and not me?" It's a guilt that festers not from wrongdoing, but from the unbearable randomness of tragedy, the knowledge that but for a different assignment, a different moment, he too could have been among the fallen. This struggle is the universal voice of those left behind. It is the insidious echo of a past that refuses to stay buried, a constant reminder of a fate avoided, and comrades lost.

Processing this trauma is not about escaping the past; it is about turning the ghosts of memory into honored guests, transforming guilt into gratitude, and ensuring that sacrifice is never endured in vain. It's a conscious, arduous process of shifting the narrative from a feeling of being underserving to a sense of profound responsibility. He realized that to truly honor them, he could not allow their memory to be diminished by his own suffering. Instead, he chose to carry their names, not as a heavy shroud, but as a vibrant banner. This transformation, profound and deeply personal, leads him to a larger purpose: inviting the nation to witness.

Corbin believes it is crucial for Americans to hear these stories; to understand the depth of the bonds forged under fire, the anguish endured, and the unwavering dedication of Marines who continue their duty even when their hearts are shattered. His journey, from the moment of shock delivered, to the agonizing wait with a rifle in his hand, to the lingering questions of why he survived, is a sobering reminder of the human cost of war. It reminds us that the battle does not end when the troops return home.

For the survivor, the battle simply changes terrain, moving from the unforgiving dust of a foreign land to the silent, invisible landscape of the heart.

Like all those who have mastered the art of compartmentalization to survive the unbearable, Corbin now seeks to slowly, deliberately unburden himself. He extends an open hand, inviting the nation to stand witness not only to the heroism of the fallen, but to the struggle of those who must live, remember, and continue to serve. This act of witness is a profound societal exchange. For the individual, it offers solace, a lessening of the isolating weight of grief that, when shared, somehow becomes lighter. For the community, it fosters a deeper understanding of the sacrifices made in its name, creating a bridge of empathy between those who served and those they served.

When a nation bears witness, it does more than just remember names or dates. It actively participates in the ongoing story of heroism and heartache. It cultivates a sense of shared human-ity, acknowledging that the invisible wounds of war are as real, and often as devastating, as the visible ones. It ensures that the bonds forged under fire are honored not just in memory, but in the sustained care and support offered to the families of the fallen and the survivors who carry their legacy. For Corbin, and countless others like him, bearing witness is the ultimate act of respect, transforming personal anguish into a shared narrative of resilience, gratitude, and an unwavering commitment to never forget. It is through this sacred act of witness that we honor the past, heal the present, and build a future where no sacrifice is ever endured in vain.

Chapter Eight

Roll Call of Remembrance

Corbin's returning home was not to a hero's welcome, but to a void. The well-meaning platitudes, "Time heals all wounds," "You must move on", felt like a betrayal. How could he move on when part of him was anchored there, amidst the men who hadn't come back?

The act of healing did not begin with forgetting. It began, and continues, with a deliberate, sacred invocation. It begins by speaking their names.

This is not a passive recollection. It is a proactive, reverent act of remembrance, where pain is turned into purpose. It is Corbin's defiance against the erasure of time, a refusal to let vibrant, complicated lives be reduced to statistics, to a line in a casualty report. Each name spoken, each anecdote shared, is a brick laid in the foundation of an enduring memorial. It is not built of stone or bronze, but of living, breathing memory. It is a roll call answered not with "here," but with "remembered."

Corbin starts, as he often does, with the youngest. Robert Hendriks. Just twenty-five. He remembers Robert as a fresh-faced, a vessel of boundless, almost exhausting energy that seemed to defy the grim reality surrounding them. He possessed an earnestness that was both charming and disarming; he believed in the mission, in the unit, in the goodness of people, with a conviction that felt both ancient and brand new. The Marine Corps, in its delayed wisdom, posthumously promoted him to Sergeant. The honor, though deserved, is a ghost of a future stolen.

Remembering SGT Robert Hendriks

But the heartbreak of Robert's story is doubled, folded over on itself in a cruel twist of fate. Roberts younger brother, Joseph, arriving in Afghanistan for his own deployment was given the devastating, sacred task of escorting his older brother's casket home. Corbin's voice grows thick when he speaks of this. He reflects on the sheer, brutal irony—the threads of life and death woven into a knot so cruel it takes your breath away. From that agony emerged a profound testament to strength. Joseph, after laying his brother to rest, returned to complete his own tour of duty. It was a quiet, staggering nod to Robert's legacy, a family's unwavering commitment echoing across a generation. In Joseph's continued service, Corbin knows, Robert's spirit lived on.

Then there was Ben Hines. Where Robert was all youthful potential, Ben was defined by a steady, dependable certainty. He was an engineer of his own future, a man who drew blueprints for his life with meticulous care. His heart was anchored firmly in Virginia, with his fiancée. Their wedding was set for October 2019. Corbin can still hear Ben's voice, buzzing with excited chatter about centerpieces and caterers, about the song for their first dance, about the house they would buy. He spoke of it with the tangible anticipation of a shared life, a future so real you could almost touch it.

Remembering SSGT Ben Hines

That future, so carefully and lovingly constructed, was snapped in an instant. The meticulous plans were replaced by a single, stark notification. The unfulfilled promise of a shared life—the dreams of a home, children, growing old together in rocking chairs on a porch they'd built—haunts Corbin. It is a searing reminder that the cost of service is pervasive, paid not only in blood on foreign soil but in the silent, aching absences in homes across the world. Corbin often thinks of Ben's fiancée, of the phantom life she must navigate, of the wedding dress that became a relic. Ben's story is a stark illustration of the ripple effect, a stone dropped in a pond, the waves of grief touching shores far from the point of impact.

And finally, there was Chris Slutman, the team's elder statesman at forty-three. Chris was different. His was not the raw energy of youth, but the grounded strength of experience. A dedicated New York City firefighter, a veteran of Ladder 27 in the Bronx, Chris had already answered a lifetime of calls. He didn't enlist until he was thirty-one, driven by a deep, passionate call to military service that superseded the hard-won security of his civilian life. It was a choice that spoke volumes about his character—a profound commitment to serving something larger than himself, even at an age when many men are settling comfortably into their roles.

Remembering SSGT Chris Slutman

He left behind his wife and their three young daughters. Corbin remembers Chris's calm demeanor, the way his experience offered a steady anchor to the younger, more volatile men. He loved to recount Chris's stories from the firehouse, a different kind of service but built on the same bedrock of camaraderie and purpose. Chris's sacrifice was layered, complex. He chose to leave a family, an established career of honor, to answer a call that resonated from a place deeper than duty. His was a courage found not in the unthinking fervor of youth, but in the mature, unwavering conviction of a man who knew exactly what he was leaving behind, and why he had to go.

Corbin keeps their memories alive. He delves into their personalities, their quirks, the stupid inside jokes that would mean nothing to anyone else but that once made them snort with laughter in the oppressive heat. He recalls the moments of shared vulnerability during mail call, the fierce, unspoken strength that forged them from a group of individuals into a single, breathing unit.

"This isn't about dwelling in the past," Corbin explains, his voice steady with hard-won clarity. "It's about keeping them present. Thinking about those times, talking about the team, sharing their experiences... it's not suppression, it's therapy. It's how I keep them alive."

A particular song on the radio can instantly dissolve the present, transporting him back to the cramped confines of a MaxxPro MRAP[1], the hum of the engine a baseline to their camaraderie. The scent of diesel on a hot day can be a punch to the gut, followed immediately by the warmth of connection. These sensory triggers, though sharp with nostalgia and sor-

---

1. **Credit DOD Programs: Mine Resistant Ambush Protected (MRAP) Family of Vehicles** MRAP is a family of vehicles designed to provide increased crew protection and vehicle survivability against current battlefield threats, such as IEDs, mines, and small arms.

row, are not avoided. They are embraced. They are the threads that tether him to the bond they shared, a bond that proved stronger than death itself.

MaxxPro MRAP

In these moments, Corbin is not just remembering; he is an architect, actively rebuilding the world they inhabited together, one cherished, aching detail at a time. He transforms the void of absence into a sanctuary filled with their enduring presence.

Through this sacred, daily roll call, Corbin is slowly unburdening himself. He is not shedding the weight by forgetting, but by carrying it more purposefully. His survivor's guilt is not erased; it transforms remembrance into a powerful, driving sense of duty. He understands, now, that the greatest honor he can offer Robert, Ben, and Chris is not to wallow in the 'what-ifs,' but to live a life worthy of the second chance they bestowed upon him.

Robert's earnestness reminds him to find joy in the mundane. Ben's dedication inspires him to build a future with intention. Chris's unwavering purpose guides him to serve in his own ways, every day.

Corbin's journey is a poignant, powerful reminder that in the profound silence of loss, love and remembrance can forge an unbreakable chain. It links the living to the lost, and through the simple, defiant act of speaking a name, transforms the cold weight of sorrow into the enduring, grateful warmth of a legacy kept alive.

In a world that often feels like a relentless tide of noise, obligation, and the insistent ping of digital demands, the very notion of a "quiet moment" can seem like an almost audacious indulgence. We tell ourselves we can't afford the luxury of pausing, of stepping away from the clamor that shapes our days. As the profound journey of Corbin so powerfully illustrates, it is precisely within these hushed interludes that life's most potent truths unfurl. It is here, in the stillness, that the soul finds its bearings, and where wisdom, so often overlooked amidst the din, whispers its most vital lessons. These aren't mere pauses; they are sacred acts of engagement, profound spaces where healing begins and purpose takes root, often in unforeseen ways.

Corbin's path to understanding this wisdom was not one paved with ease but rather forged in the trial of immense loss and unspoken pain. The weight of unseen burdens can be a crushing force, a silent battle waged in the quiet corners of the mind. The searing pain of past trauma, compounded by the devastating loss of his friends, Robert, Ben, and Chris, became an inescapable entanglement. Like so many, he hadn't always possessed the language to admit he wasn't okay, let alone the courage to ask for help. As the truth of his experience often reveals, trauma rarely arrives in isolation; "When trauma happens, it is not just the current trauma you have to deal with, but the previous things you didn't want to deal with are also thrown into the mix." This layered suffering created a formidable barrier, trapping him in a labyrinth of grief and guilt, a place where silence offered no solace, only amplification of his internal turmoil.

Chapter Nine

PTSD Takes Hold

The journey toward opening up about Post-Traumatic Stress Disorder (PTSD), for Corbin, wasn't a linear decision but a profound, almost fated, unfolding of circumstances, deeply intertwined with his burgeoning relationship with Katrina.

Corbin and Katrina's first date occurred just before Corbin's deployment to Afghanistan. Their relationship blossomed amidst the unique and often challenging constraints of military life. This was Katrina's first exposure to military life, and she was quickly immersed in its "high dose of reality." During Katrina's lunch breaks, they communicated via video calls, navigating the time zone difference as Corbin headed to bed. While Katrina experienced some anxiety about Corbin's deployment, she adapted swiftly, keeping herself busy with her work and finding solace in the support of her church, family, and friends. Katrina observed the military's advice to avoid constant news consumption, recognizing that it could cause unnecessary stress.

In her mind, she contemplated the possibility of getting married but wondered how Corbin's deployment might have changed him. Would the memories of their first date be strong enough to sustain their long-distance relationship until he returned home? And then there was the tragic event on April 8. The constant wondering how Corbin was holding up. Was he being strong in the corresponding letters and other communications to keep her from worrying, but deep-down struggling?

He is Coming Home

Then, the call came. The signal that the deployment was over, the homecoming imminent. Anticipation thrummed with a dizzying blend of elation and trepidation. Katrina wondered if Corbin would be the same person she knew and adored? When he finally returned, the landscape of their relationship was a difficult one to traverse. The war's echoes had followed him. The man who stepped off the plane wasn't entirely the man who had left. His easy laughter was a rarer commodity, replaced by a quiet, almost unnerving intensity. His gaze often drifting, as if seeing specters only he could perceive. Sleep offered little respite, his eyes hollowed by a battle unseen. Katrina soon realized that while his physical return was a profound gift, his mind remained tethered to the battlefield. The "welcome home" party, meant to celebrate his return, felt strangely muted. Friends and family offered congratulations and shared memories, but Corbin often seemed adrift, lost in the labyrinth of his own experiences.

Corbin told me that there were many "rough feelings" interwoven from his emotional distance, sudden outbursts, haunted silences, and Katrina's struggle to understand the invisible wounds he carried. They were processing raw emotions, individually and as a couple, all while simultaneously trying to build a foundation for their life together.

The deepest ache for Katrina was witnessing his internal struggle, his survivor's guilt amplified by the insidious grip of Post-Traumatic Stress Disorder (PTSD). He spoke sparingly, if at all, of the sights and sounds that haunted him. He replayed scenarios relentlessly, tormented by the question of why he was safe when three of his team were killed in action. The weight of the fallen, the unbearable loss of friends, pressed down on him, an invisible, crushing burden. This was a new, uncharted territory for Katrina.

She immersed herself in understanding PTSD, recognizing that the invisible wounds were as real, and often more devastating, than any physical injury. She became Corbin's fiercest advocate, gently encouraging him towards professional help, often sitting by his side in therapy sessions, and offering a constant, unwavering wellspring of support. This battle for his healing brought its own unique challenges for Katrina. It was a slow, arduous journey, marked by days of profound frustration, where the chasm between the Corbin she remembered and the Corbin before her felt insurmountable. There were moments of doubt, of feeling inadequate, of questioning if she was enough. But through it all, her deep love for him, and her unshakeable belief in his inherent resilience, served as her anchor.

Katrina's story is not an isolated one. It is a powerful testament to the extraordinary strength and unwavering sacrifice of military families – the silent heroes who hold the fort, shoulder anxieties, and then navigate the complex terrain of homecoming. The physical return of a service member is merely the prologue to a new chapter, one that demands a different kind of courage, a deeper well of understanding, and a profound commitment to healing, united. The echoes of war may persist, but with enduring love, patient dedication, and unwavering support, Katrina and Corbin found their way back to forging a future – not free from the shadows of the past but illuminated by the indomitable light of their shared strength.

Amidst this emotional journey, an unexpected event offered a poignant turn of events. Corbin's fallen comrade, Ben, was being honored through a unique and deeply personal tribute orchestrated by his fiancée. She had established the @findingsneakywalnut Instagram page, where individuals could wear their "Ben" memorial T-shirts and tag the page, creating a virtual network of remembrance and shared grief. This heartfelt project was intended to preserve Ben's memory with a concept that Ben was "seen" everywhere. The T-shirts simply stated the date Ben was killed covered by a black ribbon.

T-Shirt remembering Ben's sacrifice

Corbin and Katrina embarked on a journey to Caledonia Park for an outdoor market event. Afterward, they had planned to visit the Gettysburg Battlefield to capture photographs of their "Ben" shirts and his fiancée on their location. They also intended to spend some time hiking. As they were on their way to Gettysburg, Corbin and Katrina stopped at Caledonia. They were already wearing their Ben shirts when they strolled around the various vendor tents. During their walk, they encountered a gentleman named Bud, who noticed Corbin's shirt and the KIA ribbon prominently displayed on the front.

A fellow veteran, Bud Wagner, spotted Corbin across the crowd. Perhaps it was the t-shirt, or perhaps Bud, with a veteran's unique eye, recognized fragments of his own journey in Corbin's guarded pain. Bud approached, his voice gentle and non-judgmental, asking Corbin simply, "Are you okay?" This immediate connection, born from shared experience and mutual respect, opened a door. They continued talking, and it became clear that Bud was part of the [1] REBOOT program, an organization dedicated to helping veterans heal from trauma. In a move uncharacteristic for him, Bud happened to have a REBOOT business card in his pocket that day, which he offered to Corbin, urging him to consider following up.

---

1. **Military REBOOT** is a 12-week, faith-based, peer-led course that helps veterans, active-duty military and their families heal from service-related trauma.At REBOOT courses across the country, military families are reconciling, divorce rates are dropping, medication abuse is decreasing, and suicide numbers are falling. Credit: http s://rebootrecovery.com

Bud and Melinda Wagner, Corbin and Katrina Sipe

It was Katrina who seized this opportunity to follow up with Bud with unwavering determination. Recognizing the glimmer of hope it offered, she reached out to Bud on social media and persisted with Bud to secure an early meeting. Despite a typically long waiting list, her insistence, coupled with Bud's endorsement and perhaps a sense of urgency from the RE-BOOT team, opened a path. Bud, demonstrating his deep commitment to helping a fellow veteran, agreed to meet them at a location halfway between their homes, further easing the initial hurdle.

They decided to give the meeting a try, and to their surprise, found themselves looking forward to the follow-up sessions. REBOOT became the first crucial, structured step in Corbin's healing journey. It was a space where his experiences were validated, and he began to process the trauma he had carried in silence for so long.

After completing the REBOOT program, Corbin understood that the work was far from over. REBOOT was the gateway that ultimately led Corbin to seek ongoing, individualized counseling and to embark on the arduous, ultimately liberating, process of confronting his past. This initial courage to acknowledge and address his inner chaos was the essential prerequisite for finding any quiet moment truly restorative. Without first facing the storm within, peace remains a distant, unattainable shore.

From this hard-won foundation of active healing, Corbin began to discover the profound power of intentional stillness. He came to understand that perhaps the ultimate act of honoring his fallen friends was to live fully, to embrace the moments they were so cruelly denied. This profound commitment began to transform the corrosive grip of survivor's guilt into a deep and active gratefulness. "Remembering that the Lord has Corbin here for a reason and to live for another day," he reflects, his voice imbued with a quiet reverence. This recognition guides his appreciation for what he holds today, propelling him toward moments of quiet contemplation, a conscious and deliberate slowing down to observe the simple, often overlooked beauty of the world around him. He finds a profound sense of grounding by "walking in nature and taking time to appreciate nature, water, flowers, etc. Things that are typically overlooked by society."

Nature's Calming Allure

In these moments, where the hurried pace of the outside world fades into a gentle hum and the spirit is finally allowed to breathe, he discovers a multifaceted connection and a sense of refinement. The pain of loss is still present, a scar etched deeply into his being, but it is now met with a quiet, resilient peace. This is the very core of his healing: the quiet acceptance that even amidst the most agonizing of circumstances, there is an enduring path forward, a path guided by an unwavering faith. Corbin told me, "As painful as it is, it's still the Lord's hand that things happen the way they did. In the Lord's wisdom, he is still with us through the painful times." These quiet moments are not about forgetting, but about integration; they are not about denying the presence of pain, but about finding solace alongside it, recognizing the enduring presence of grace woven through the fabric of his life. The wisdom found here is the profound understanding that true strength isn't found in perpetual motion, but in strategic, deliberate stillness – a stillness that allows the heart to mend and the soul to recalibrate, to find its true north once more.

Though technically holding the status of Inactive Ready Reserve, Corbin remains, in spirit and in deed, undeniably a Marine. "Once a Marine, always a Marine. Brotherhood of the Marine Core," he affirms, a testament to the enduring bond that continues to provide him with unwavering support and an irreplaceable sense of belonging. This brotherhood, a constant source of strength, is also a poignant reminder of the sacrifices made, deepening his commitment to active, present-day gratitude. In his daily life, Corbin actively nurtures this gratitude, a potent balm against the lingering shadows of trauma. He constantly revisits Bud's wisdom, remembering that his fallen team members would want him to embrace joy, to live fully and vibrantly. Whenever the Lord leads, he reaches out to Gold Star families, offering comfort, sharing cherished memories, and extending his deep compassion to others who walk a similarly painful path of grief. These acts of service are not separate from his quiet moments, but rather directly flow from them, fueled by the empathy and clarity they provide.

Corbin's ultimate message is one of unwavering faith and enduring hope. "The Lord is still with us in those extremely low places; my hope is that as a believer, others will come to know that God is still good in those terrible things." He acknowledges the profound difficulty of comprehending why certain events unfold as they do, yet trusts implicitly in God's inherent goodness, even amidst the deepest suffering. "I can't look at this as a believer and say that the Lord's hand is not in this, but I can't say God isn't good in this. Even if we don't understand why, we don't have to understand why. It's the reality of it." This profound acceptance, born in the quiet, reflective space of faith, frees him from the burden of needing all the answers, allowing him to simply be present in each moment, grateful for the gift of each breath.

His journey stands as a powerful, resonant reminder to us all: *processing trauma is not a destination to be reached, but rather a continuous act of honoring the past while dedicating oneself with fervent gratitude to the present.* By speaking the names of his fallen team members—Robert, Ben, and Chris—Corbin ensures their sacrifice continues to resonate, not as an ending, but as the enduring reason he chooses to appreciate life every single day. The wisdom of the quiet moments, then, is not merely about finding a brief, fleeting respite from the overwhelming world, but about cultivating a deeper awareness, a profound gratitude, and an unwavering commitment to a life lived with purpose and boundless compassion. It is in these sacred moments of stillness that we truly hear the quiet, insistent call to live fully, to heal authentically, and to honor the precious, irreplaceable gift of existence.

The echo of a fallen hero's final breath doesn't always resonate in thunderous pronouncements or the crisp formality of a military send-off. While these moments acknowledge the ultimate sacrifice, the true, enduring measure of that cost is often found in the quiet chambers of the heart, in the altered rhythms of daily life, and in the profound, often unseen, adjustments made by those left behind. For Gold Star families, the indelible mark of loss is etched not in public fanfare, but in the tender, persistent ache of absence.

Chapter Ten

The Quiet Strength of Purpose

Healing both the Body and Mind

Life often presents us with paths that feel less like wide avenues and more like treacherous, winding trails. We navigate these passages searching for glimmer of light, a purpose to anchor ourselves against the prevailing winds.

The specific contours of Corbin's path are his alone. What is clear is that he faced his own difficult chapters, moments when his path forward seemed obscured, and the very concept of "solace" felt like an unreachable dream. It was through the rigorous discipline of fitness, the tangible act of moving his body, pushing its limits, and demanding more of himself, that he began to piece himself back together. It wasn't just physical strength he sought; it was a lifeline, a way to mend what felt irrevocably broken inside. The iron became his anchor. The sweat, the strain, the singular focus required for each rep – these became his meditation, his therapy, a way to silence the clamor of his own mind and find a new kind of peace in the stark, simple truth of effort.

That personal journey, often solitary and arduous, illuminated a powerful truth for Corbin: the path to wellness, to wholeness, often lies directly through discomfort. It's a narrow gate, indeed, demanding courage and perseverance. And, once he found his footing, once he felt the transformative power of that journey within himself, he knew he couldn't keep it to himself.

Narrow Gate Strength and Conditioning was born from this deep-seated, empathetic need to share, to create a space where others could experience their own profound metamorphosis.

Inspired by colleagues who recognized his innate passion for fitness, Corbin brought Narrow Gate Strength and Conditioning to life. Rooted in the unwavering discipline of his military experience, his philosophy emphasizes the symbiotic and indispensable relationship between physical prowess and a strong, resilient mind. The gym's name and it mission draw profound inspiration from Matthew 7:13-14 (NIV): "Enter through the narrow gate. For wide is the gate and broad is the road that leads to destruction, and many enter through it. But small is the gate and narrow the road that leads to life, and only a few find it." This scripture perfectly encapsulates the dedication and intentionality required for true strength and holistic well-being. "Conditioning takes discipline," Corbin explains. "In our society, there are many things available to participate in that are not healthy. So, training takes discipline; our bodies are conditioned to do things differently, all the benefits of training: heart pumping more blood efficiently, growing through and processing the issues, now having a greater appreciation of the body and capabilities that the bodies are made to do." He infuses his training with the deep understanding that mental fortitude is crucial, drawing poignant parallels between the challenges faced in rigorous physical training and the formidable obstacles overcome in life. The narrow gate, for Corbin, is also about the disciplined choice to seek out quiet moments, to process, to heal, and to consciously choose to live a life imbued with intentionality and profound gratitude.

The name itself is a quiet manifesto. It speaks not of ease or open fields, but of a deliberate, singular passage. The true beauty of Narrow Gate Strength and Conditioning lies in this profound understanding of human potential. It acknowledges that growth is often uncomfortable, that true strength is built with the challenge, and that the most rewarding journeys are rarely the easiest. Corbin, with his quiet intensity and deep empathy, embodies this philosophy. He creates a space where clients can confront their limitations, not with judgement or fear, but with curiosity and a growing sense of self-efficacy. It's a place where the act of pushing through the narrow gate becomes not just a physical feat, but a profound transformation – journey toward a stronger body, a more resilient mind, and a life lived with greater confidence and purpose.

For many, teaching is a job. For Corbin, it is a profound act of self-preservation, a daily reaffirmation of his own healing. Each session at Narrow Gate Strength and Conditioning isn't just about instructing; it's about connecting, witnessing, and guiding. When he walks among his clients, his eyes keenly observing their form, his voice calm but firm, he sees more than just muscle and bone. He sees in every hesitant glance, every strained grimace, every moment of self-doubt, a reflection of his own past struggles. This isn't theoretical empathy; its empathy developed through personal experience.

His solace as an instructor doesn't come from the applause or the accolades. It's a quieter, more internal process. It comes from the almost imperceptible shift in a client's posture as they finally find their core strength. It comes from the flicker of determination in someone's eyes who, just weeks ago, thought they couldn't possibly lift that weight or hold that plank for another second. It comes from the shared, unspoken understanding when a challenging set is completed, and a collective exhale sweeps through the room, punctuated by his quiet, affirming nod. "Good. Now we breathe. Now we reset."

He remembers what it felt like to be lost, to feel physically and emotionally adrift. And so, he doesn't just bark orders; he offers encouragement born of lived experience. He doesn't just push; he pulls, reminding everyone of the strength they already possess, even when they can't quite see it themselves. When a client expresses frustration, Corbin doesn't dismiss it. He listens, acknowledges the difficulty, and then, with a quiet resolve, helps them find the next small step forward. "It's not about the weight today," he might say. "It's about the intent. It's about showing up." This process of guiding others through their own narrow gate is, for him, a constant reminder of his own triumph, a continuous loop of giving and receiving strength.

The community at Narrow Gate Strength and Conditioning is another profound wellspring of Corbin's solace. He has meticulously cultivated an environment that is challenging and supportive. It's a place where competition is secondary to collective growth, where personal bests are celebrated by everyone, and where vulnerabilities are met with understanding, not judgment. In this shared space, Corbin sees individuals not only transforming their bodies but also their minds. He witnesses hesitant smiles turn into confident grins, quiet whispers become firm declarations of "I can." The energy of this collective journey—the shared sweat, the mutual encouragement, the quiet camaraderie—is palpable, and Corbin is at its epicenter, his own spirit nourished by the strength he helps to build around him.

Consider Sarah (not her real name), who arrived at Narrow Gate Strength and Conditioning burdened by chronic stress and a sense of physical stagnation. Initially, every movement was a struggle, her confidence as fragile as glass. Corbin, with his characteristic patience, didn't push her beyond her limits but consistently encouraged her to explore them. He showed her modifications, celebrated tiny victories like holding a plank for five full seconds, and reminded her that consistency, not perfection, was the key. Watching Sarah, months later, move with a newfound grace and power, her eyes shining with a quiet pride as she deadlifts a weight, she once thought impossible, is a profound reinforcement for Corbin. It's a tangible manifestation of the very healing he once sought for himself. Her journey, like so many others, becomes a mirror reflecting his own enduring purpose.

The irony, perhaps, is that in seeking to guide others, Corbin continually guides himself. Each class he instructs, each life he touches, reaffirms the power of the path he chose. The solace isn't a destination he reached long ago; it's a dynamic, living force, continually renewed by the courage and transformation he witnesses every day within the walls of Narrow Gate Strength and Conditioning. He has built more than a fitness company; he has built a sanctuary, a testament to resilience, and a quiet, powerful echo of his own journey from shadow into light.

It's not just about building muscle or mastering a complex lift; it's about forging a resilient spirit, a body capable of navigating life's challenges, and a confidence that echoes far beyond the gym walls. He sees the potential within everyone, the dormant strength waiting to be awakened, and he recognizes that the path to unlocking it is rarely a straight, unblemished line. It's often a winding trail, fraught with self-doubt and the seductive siren song of the easy way out.

This understanding informs everything. He doesn't simply prescribe exercises; he partners with his clients, offering a holistic approach that acknowledges the interconnectedness of physical prowess, nutritional well-being, and mental fortitude. "This isn't just about workouts," he emphasizes, his words carrying the quiet conviction of lived experience. "I'll help you get your nutrition in order, build habits that actually stick, and stay on track with your goals. I'll push you where you need it and keep you moving forward. Every session will be intentional; every effort will matter."

In a world saturated with quick fixes and superficial transformations, he champions a philosophy built on consistency, intelligent coaching, and a deep understanding of what it takes to achieve sustainable results. His approach is rooted in the belief that true progress, the kind that rebuilds you from the inside out, requires a steady hand and a focused mind.

And in that deliberate, often challenging, act of squeezing through, he, and those he guides, discover a strength they never knew they possessed—a strength that opens a broader, more vibrant horizon, together. For Corbin, the narrow gate was never an obstacle; it was the way through. And in holding it open for others; he found his way home.

Where Discipline Meets Transformation

So, what are the key tenets at Narrow Gate Strength and Conditioning? The disciplines of Olympic weightlifting, Brazilian Jiu-Jitsu, and Muay Thai are more than just workouts—they are portals into deeper layers of human potential. Each art offers a distinct challenge, a language of movement that teaches us not only about the body but also about resilience, patience, and the quiet power of persistence. This isn't just training; it's transformation, sculpted in the deliberate practice and guided by an ethos that prioritizes both effort and empathy.

[1] Pillars of Power and Perseverance

Olympic Weightlifting: There is something elemental about lifting immense weight off the ground, defying gravity, if only for a moment. The clean & jerk, the snatch, they are feats of both explosive power and delicate precision. A barbell loaded with iron doesn't care about your excuses. It demands total engagement, core braced, breath controlled, mind sharpened to a fine point. The lift is a conversation between body and physics, where hesitation means failure, and commitment means transcendence.

---

1. Descriptions of the different types of training are taken directly from the Narrow Gate and Strength and Conditioning website: https://narrowgatestrength.com

Weightlifting teaches discipline in its purest form. It reveals how small technical adjustments, the angle of a wrist, the timing of a hip thrust, can mean the difference between struggle and success. It's a metaphor for life: the way we must sometimes summon all our strength at once, channeling controlled aggression to move what once seemed unmovable.

But beyond the raw physicality, weightlifting is a lesson in humility. The barbell humbles everyone eventually. Plateaus (where you feel stuck despite your effort) arrive without warning. A lift that felt effortless yesterday suddenly feels impossible today. And that's where the real work begins, showing up anyway, refining form, trusting the process, and understanding that progress is rarely linear.

Brazilian Jiu-Jitsu (BJJ): If weightlifting is about domination through force, BJJ is the art of finesse under pressure. Its chess played with the body. A smaller, weaker opponent can neutralize a larger one through leverage, timing, and strategy. There's a reason it's called the gentle art; it proves that strength isn't always about overpowering; sometimes, it's about outsmarting.

On the mats, you learn the weight of another person's will. You learn to breathe when someone is trying to choke you. You learn that panic is the real enemy, not the opponent. BJJ reveals how often we react instead of responding, how our instincts, untrained, can lead us into worse positions. But with practice, the flinch reflex fades. You start to see openings where you once only saw threats.

And then there's the surrender—the tap. In a culture obsessed with domination, BJJ teaches the wisdom of knowing when you're beaten. There's no shame in tapping; it's an act of self-preservation and learning. The lesson? Sometimes, yielding is the smartest way to survive long enough to fight another day.

Muay Thai: Known as the art of eight limbs (fists, elbows, knees, shins), Muay Thai is unforgiving. It tests your endurance, your pain tolerance, your ability to keep moving when your lungs burn and your muscles scream. Every strike is an exercise in commitment—half-hearted technique gets punished. You learn to take hits as much as you learn to deliver them.

There's a raw honesty in Muay Thai. No flashy tricks, just efficient brutality. But beneath the toughness, there's also profound discipline. Control. Timing. Distance management. A skilled Nak Muay (practitioner) knows that power without strategy is wasted energy. And perhaps most importantly, Muay Thai teaches you to stay composed in chaos. When fatigue sets in, when adrenaline blurs your focus, can you still breathe? Still think? Still move with intention?

The Coach Who Sees More Than Muscles

Corbin's presence is the keystone of Narrow Gate Strength and Conditioning. He doesn't just instruct; he perceives. Training isn't just about sets and reps; it's about the person behind the effort. He recognizes the hesitation in a first-time lifter's stance, the fleeting doubt before stepping onto the mats, the way someone's jaw tightens when they're pushing through a grueling round on the heavy bag.

His empathy allows him to meet people where they are, whether they're elite athletes or complete beginners. He knows that fear and frustration are part of the journey. The woman who hesitates before attempting a PR deadlift isn't just battling the weight; she's battling the voice that whispers, "What if I fail?" The man struggling to escape side control in BJJ isn't just stuck physically; he's confronting the helplessness we all feel in life's suffocating moments.

Corbin's gift is his ability to tailor the challenge. He knows when to push "One more round. You've got this." and when to pull back "Let's dial it down and focus on technique." His quiet intensity isn't intimidating, it's reassuring. It says, "I believe in you, even when you don't."

Beyond the Gym: The Science of Sustainability

No one thrives on willpower alone. Corbin understands that discipline in the gym is unsustainable without discipline in the kitchen, in sleep, and in stress management. The body isn't a machine; it's an ecosystem. What you feed it, how you rest it, and how you treat it daily determines whether your efforts yield progress or burnout.

His nutritional coaching isn't about rigid meal plans or demonizing foods. It's about awareness, understanding how different fuels affect performance, recovery, and mood. He helps clients build habits, not restrictions. Because progress isn't about perfection; it's about consistency amid imperfection.

The Narrow Gate Strength and Conditioning Philosophy

The name isn't arbitrary. The "narrow gate" is a metaphor for the challenging, intentional path—the one that demands more but rewards deeply. In Matthew 7:13-14 (the biblical reference), it's described as the road less traveled, the difficult route that leads to life.

It's not about mindless suffering or brute-force effort. It's about purposeful struggle. It's the understanding that real growth happens in the discomfort zone, that moment in a heavy squat where every fiber screams to quit... and you don't. The moment in sparring where your instincts say run... and you stay. The moment when old habits tempt you... and you choose better.

This is where Corbin's "push hard, focus on results" mantra takes root. Not as a blind grind, but as a commitment to showing up, for the hard days, the frustrating plateaus, the small wins that no one else sees.

The Unseen Ascent

Strength isn't built in highlight reels. It's forged in the quiet hours, the extra mobility work after class, the meal prepped when exhausted, the cold morning when the bed is so much more appealing than the gym. The world sees the result, the lifted weight, the submission won, the fight endured, but the true journey is invisible.

Training

Corbin isn't just a coach; he's a guide for that journey. He doesn't carry you up the mountain. He points to the path, hands you the tools, and walks beside you. The rest? That's on you. And that's the beauty of it.

Because when you walk through the narrow gate, you don't just emerge stronger. You emerge changed.

Chapter Eleven

Legacies That Whisper Hope – Gold Star Families

The world often seems deaf to the silent screams of a wounded heart. Grief, in its rawest form, is profoundly isolating. Military grief carries an additional, unique weight. It's a loss rooted in sacrifice, in duty, in service to something larger than oneself, which can make it even harder for those outside the military community to truly grasp. The hero's departure leaves behind not just an absence, but a legacy, a complex mix whether of pride, pain, and unfulfilled futures. The survivor, often a spouse, parent, child, sibling, or fiancée, is left to navigate this intricate landscape alone, often feeling invisible amidst the visible world. The "move on" platitudes, however well-intentioned, only deepen the chasm, suggesting that their pain is an inconvenience, rather than a sacred, necessary process. It is in this overwhelming noise and profound solitude that the search for stillness, for a space where the heart can finally breathe, acknowledge its wounds, and begin the arduous work of healing, becomes an urgent, existential quest.

Gold Star families, those who have endured the unimaginable sorrow of losing an immediate family member in military service to their nation, carry a unique and profound weight. Their sacrifice is not a singular event, but a lifecycle of adaptation and resilience. They possess a quiet, hard-won wisdom, gleaned from navigating the labyrinth of grief, loss, and the often-unforeseen complexities that follow such a profound rupture. This wisdom speaks to the fragility of our existence, the depth of unwavering commitment, and the fundamental human need for a steadfast network of support.

Corbin's personal journey of healing and transformation has ignited within him a fierce and compassionate passion: advocating for Gold Star families. He understands, with a depth born from his own experience, that "It's our responsibility as Americans not only to honor the service of their loved ones who gave their lives for the country, but to also recognize the toll on families and their roles in that as well. It's not just a name, not just this Marine was killed, the families are tremendously affected in their loved ones' death as well." The quiet moments of reflection on his own pain and subsequent transformation have allowed him to empathize profoundly with the unseen struggles of these families, struggles that often go unacknowledged and unaddressed. He believes we fall tragically short in meeting their profound needs, particularly in the realm of mental health. The stark and sobering reality that veteran suicides far surpass those killed in action underscores the urgency and critical importance of his message.

"Your mental health is important, having mental health challenges is not anything to be ashamed of. Your mental wellbeing is a part of your overall wellbeing, part of your overall health,"

Corbin emphasizes, his words carrying the weight of experience and the resonance of a lifeline for those battling in silence. His advocacy is a powerful testament to how the wisdom gleaned in quiet personal reflection can ignite a powerful public mission, transforming individual suffering into a catalyst for collective healing and understanding.

Fortunately, in communities across the nation, a network of dedicated organizations operate with quiet tenacity and profound purpose. These Gold Star foundations are the silent guardians, the unseen hands that reach out, offering solace, stability, and practical assistance. They understand that the sacrifice of our heroes is an enduring legacy that demands not only remembrance but also active support for those who carry the deepest imprint of that loss. These foundations do not possess the power to erase the pain of grief, nor do they claim to. Instead, they provide something equally vital: the stable ground upon which the arduous journey of healing can begin. They weave threads of hope into the fabric of despair, transforming those crushing quiet moments of solitude into opportunities for growth, resilience, and ultimately, grace.

It is within this context that the tireless advocacy of these types of organizations is not just important, but essential. By lending their voice and energy to shed light on the realities faced by Gold Star families, they serve as vital conduits, reminding us that true honor for the fallen is incomplete without an unwavering dedication to the living. These families, already grappling with the overwhelming tide of grief, frequently confront a bewildering array of practical and emotional obstacles. Financial instability can emerge like a phantom, leaving them vulnerable. Logistical challenges, from managing affairs to simply navigating bureaucratic systems, can feel like insurmountable mountains. And perhaps most debilitating of all is the chilling embrace of emotional isolation, a void so vast that even the seemingly simple act of drawing breath can feel monumentally difficult.

The silence that falls after a loved one's final salute is not empty; it is a deafening void, a sudden chasm where love, laughter, and a future once resided. For military families, the ground beneath their feet doesn't just shake when a service member falls; it crumbles, leaving them teetering on the precipice of an unimaginable new reality.

Chapter Twelve

The Continual Journey

Importance of Companionship

Life, in its breathtaking beauty and bewildering complexity, often asks more of us than we feel capable of giving. This reality defines the human condition: a dynamic tension between the soaring peaks of joy, discovery, and connection, and the unpredictable, often overwhelming variables of existence. We are constantly navigating shifting ethical demands, relational complexities, and the brute force of random misfortune. There are indeed seasons of sunshine that lift our spirits effortlessly, filling our sails with boundless energy and optimism. And then there are the longer, darker stretches, vast emotional winters when the weight of the world seems to settle squarely on our shoulders, a crushing burden composed of global anxieties, personal grief, economic strain, and the silent, corrosive pressure of unmet expectations.

In these moments, when the path ahead feels obscured by mist and exhaustion, and the heart grows heavy, ancient wisdom whispers to us. These are not ephemeral voices, but the distilled knowledge of generations who have faced similar trials, offering gentle reminders that we are neither truly alone in our suffering, nor are we without light. This inherited resilience points away from isolated self-reliance and toward community. These are not grand pronouncements cast down from on high, demanding perfection or instant recovery, but rather compassionate invitations to pause, to look inward and outward with honest assessment, and to lean into a strength far greater than our own singular will, a communal, historical strength.

The Radical Act of Vulnerability

The first crucial step in accepting this invitation requires an act of genuine, radical vulnerability:

Have the courage to ask for help and cease the exhausting performance of maintaining a facade that everything is okay.

The pressure to appear perpetually competent, resilient, and cheerful is one of the most isolating forces of modern life. That polished exterior, that automatic "I'm fine" delivered through gritted teeth, is not only a lie to those around you, but a profound betrayal of your present emotional needs. True strength is not the ability to bear every load alone until you collapse; it is the wisdom to delegate, to share the weight, and to admit, "I cannot manage this right now." When we drop the facade, we do not surrender; we open a channel for genuine connection, allowing others who love us the dignity of being truly helpful. This openness transforms a private crisis into a shared, manageable challenge.

The Profound Truth of Stillness

Beyond seeking external support, one of the most profound truths we can embrace is the power of reflection. In our fast-paced world, where demands are relentless, digital notifications constant, and the imperative to do outweighs the necessity of being, we rarely grant ourselves the quiet, sacred space required for genuine introspection. We rush past the markers of stress and trauma, never processing the emotional data.

It is in this enforced stillness; the deliberate downtime, the journal entry, the walk without music, that clarity often emerges. Reflection acts as an anchor in the storm; it organizes the chaos of the mind, allowing the sediment of anxiety and distraction to settle. When we pause, we can measure the true dimensions of our burdens, distinguish between genuine crisis and manufactured urgency, and, most importantly, identify the next, smallest, necessary step forward. This deliberate act of looking inward does not solve the problem instantly, but it ensures that when we finally move, we do so from a place of considered wisdom, resting not on false bravado, but on the enduring strength discovered in the quiet depths of the self.

Take time, truly take time, to reflect on the nature around you.

Step outside, wherever you are. It doesn't have to be a majestic mountain vista; it can be the persistent green shoot pushing through a crack in the pavement, the intricate pattern of a leaf vein, the way sunlight filters through the branches of a tree, or the rhythmic murmur of a nearby stream. Observe the unwavering resilience of nature, its inherent beauty, its ceaseless cycle of growth, decay, and renewal.

Leisurely Walk Through the Woods

In the rustle of leaves, hear the voice of enduring patience. In the vastness of the sky, find a sense of perspective that shrinks your immediate worries. The unyielding strength of an ancient tree, standing firm through seasons of storm and calm, can be a silent testament to your own inner fortitude. Nature doesn't rush, doesn't fret about tomorrow; it simply is, embracing each moment with an innate grace. When we allow ourselves to be immersed in this simple, profound presence, a sense of calm takes root within us. It reminds us that we are part of something much larger, a beautiful, elaborate masterpiece where every thread, including our own, holds an essential place. This connection to the natural world is a balm for the weary soul, a gentle anchoring during life's swirling currents. It is a moment to breathe, to remember the inherent goodness and order that persists, even when our personal worlds feel chaotic.

This reflection often brings us face to face with another undeniable truth: you don't know what tomorrow holds.

This reality can be a source of profound anxiety to plan every conceivable outcome. But it can also be a liberating whisper, an invitation to release the suffocating grip of future worries and embrace the sacredness of the present moment. We spend so much energy on what might be, that we often miss the gifts of what is. The truth is that tomorrow is a page still to be written. While wise planning is prudent, obsessive control is futile. Instead, consider allowing this uncertainty to foster a deeper sense of trust; trust in life's unfolding resilience you've already demonstrated, and trust in a divine plan that transcends your own limited understanding. Living with an open heart to tomorrow's unknown allows us to fully inhabit today, to savor its joys, face its challenges with courage, and appreciate the fleeting beauty that surrounds us. It's a gentle surrender, not of agency, but of anxiety.

Even with the peace found in nature and the wisdom of living in the present, there will be burdens too heavy to bear alone. There are sorrows that carve deep valleys in the heart, decisions that feel impossibly complex, and moments of despair that threaten to overwhelm. In these times, the most courageous act is often the simplest: reach out to others for help.

We live in a world that often champions fierce independence, equating vulnerability with weakness. But this is a profound misunderstanding of the human spirit. To reach out, to confess your struggle, to admit your need for support, is an act of immense strength and humility. Don't carry the burden alone.

There are hearts ready to listen, shoulders ready to offer support, hands ready to help lift the weight.

Corbin emphasized that this impulse to connect, to share, to bear one another's burdens, is deeply ingrained in the human spirit and finds its most profound expression in a life committed to Christ. To commit to living the life of Christ is to embrace a path of radical love, compassion, and service. It means seeing the divine spark in every person, offering forgiveness not just when it's easy, but when it's difficult, and extending grace even when it feels undeserved. It means actively seeking justice for the marginalized, comforting the sorrowful, and lifting the downtrodden. It means cultivating a heart of empathy, stepping into another's shoes, and feeling their pain as your own, just as Christ did.

This commitment transforms how we experience every aspect of life. When you reflect on nature, you see it as a magnificent cathedral, a testament to God's boundless creativity and unwavering presence. When you consider the uncertainty of tomorrow, you do so not with fear, but with a trusting faith in a loving Providence that holds your future in its hands. And when burdens weigh heavily, you are reminded that Christ himself invited us, "Come to me, all you who are weary and burdened, and I will give you rest." (Matthew 11:28 NIV). Living the life of Christ isn't about rigid adherence to rules, but about cultivating a relationship, a profound surrender to a love that is unconditional and unending. It equips you with the strength to face tribulations, the wisdom to navigate complexities, and the humility to accept help when it's offered, both earthly and divine. It encourages you to be the hands and feet of God for others, and to gratefully receive comfort from others whom God sends to you. It teaches you that true strength lies not in never falling, but in rising again through grace, and in helping others to rise with you.

In this journey, every moment becomes an opportunity for growth and connection. The quiet contemplation of a sunrise, the honest confession of a struggle to a trusted friend, the simple act of lending a listening ear to another, and the conscious decision to walk in the footsteps of Christ – all these threads weave together to form a life of profound meaning and purpose. You are invited, not commanded, to a life of abundance, not necessarily in material wealth, but in spiritual richness, profound peace, and enduring love.

Relaxing by the Lake

So, dear heart, when the world feels too much, remember these gentle truths. Pause, breathe, and let the wisdom of nature soothe your soul. Release the grip of tomorrow's unknowns and embrace the sacred present. Never believe you must carry your burdens alone; there is strength in vulnerability, and comfort in community. And above all, consider committing your path to the life of Christ, allowing His boundless love and unwavering compassion to be your compass, your comfort, and your greatest source of strength. For in truly living His life, you discover your own, imbued with purpose, peace, and the radiant hope that transcends all earthly struggles.

Chapter Thirteen

Support for Veterans and Gold Star Families

The silence that follows a profound sacrifice is deafening. It is a charged quietude into which our nation must place its most sincere and enduring response. Our collective expression of gratitude for the sacrifices made by service members and their families—especially the ultimate sacrifice accepted by Gold Star families, must extend far beyond brief platitudes and symbolic salutes. True support is not merely loud in its generosity; it is a tangible, empathetic, and thoughtful action undertaken with measured intent.

Supporting Gold Star families is not an optional kindness; it is a collective duty, a sacred trust we hold as a nation, a covenant that demands we honor the living legacy of the fallen. To fulfill this trust requires us to move beyond instinctive, emotional reactions and engage in our own measured, quiet analysis. The most effective support is strategic in its intention and deeply knowledgeable about where the need truly lies.

This measured approach acknowledges the gravity of the gift: the donation is not simply money; it is a piece of earned trust offered in the name of a hero. To move with discernment is to honor the sacrifice with integrity. Corbin told me that when selecting a Gold Star organization to support, we are called to ensure our resources flow to those who exhibit the highest standards of stewardship and transparency. The legacies that give us hope are the ones that don't let the ultimate sacrifice end in silence. They show us that grief, even though it's a part of life, doesn't have to be the end. It can be turned into a powerful reason to act, driving work that builds and supports the very communities the heroes promised to protect.

By supporting these foundations with care, honesty, and purpose, we do more than just give money. We actively and humbly participate in the deep act of remembering. We make sure that the spirit of every hero, brings comfort to grieving families and countless others whose names are in their families' hearts—becomes the quiet force that keeps positive change going.

The ripple effect of their quiet work becomes our nation's hope: a steady, growing flow that turns sadness into service, making sure that the light of those who have passed on keeps guiding future generations. This is the lasting, caring duty we owe—to listen for the whisper of hope and join the quiet, important work it inspires.

Strategic Generosity

As Corbin and I concluded our discussions about his journey, we delved into how to support Veterans and Gold Star families. Corbin's passion was evident as he stressed the significance of mindful spending to maximize our impact. He described the vast landscape of remembrance, filled with numerous foundations and support groups, each born from profound grief and driven by a unique mission. While our hearts naturally urge us to help, Corbin said that we should use our minds and not just emotions to select the right recipients. By practicing strategic generosity, we can ensure that the funds dedicated to honoring a hero are amplified through effective and ethical management. This thoughtful approach allows us to fully embrace the mission of transforming sorrow into service, offering hope and support to those who need it most. Corbin said that

To truly make a difference, we need to focus on three key areas: being open, making a real impact, and working together.

Building Trust Through Openness

Being open means we can trust that the money we give is being used wisely and fairly for what we need it for. We should look for organizations that are upfront and honest about how they manage their money and run their programs. This means checking out their financial reports, program reports, and governance policies, not just looking at their pretty websites.

When a foundation is open, it shows they really care about the donor, the community, and the people they help. It means they're not just trying to get famous or make money for themselves, but they're really focused on helping others. Good stewardship means keeping costs down and making sure most of the money goes directly to helping families with what they need, like grief counseling, school help, or special support. This honesty is the key to building trust and making sure we can invest our trust with confidence.

Focusing on Real Change, Not Just Words

When we research an organization, we need to be careful and focused. We should look past flashy ads and emotional appeals to see what real, lasting change they're making for families. The important thing is not how much they advertise, but what specific, real services they offer and how long they can keep providing them.

True impact is seen in real, meaningful changes for people: like scholarship winners who finish school and find a stable life, veterans' spouses who get the help they need to heal after years of tough grief, and families who get a place to live when they're going through a medical crisis. When we talk about focused impact, we're talking about being mature enough to do what you set out to do well, keep track of what's working, and change your plans when you need to, not just when it's easy. We want foundations that are mature in how they run things, able to help people for a long time, even after everyone stops talking about them. They're not just making big promises; they're quietly working to rebuild lives every day.

Commitment through Resonance[1]

To keep going, you must feel it inside. Gold Star foundations help lots of different people, from getting homes and scholarships to helping with mental health and changing laws. We need to pick an organization whose mission we feel connected to, so our support is not just a one-time thing but a long-term investment.

If you're thinking about helping with immediate money problems (like the ongoing work of organizations that give homes without mortgages), making sure kids can keep going to school, or helping with the special mental health issues that surviving spouses and parents face, supporting something that means a lot to you turns a donation into a personal investment. This connection makes sure you'll stick with it, so you can help not just with money, but with your heart, knowing that your help is really making a difference in something you care about.

---

1. Information about the various organizations was collected from open access information including information from organizational websites.

While many local foundations focus on turning specific grief into community action, like making sure a fallen hero's legacy lives on in their hometown, we should also remember the powerful organizations that quietly make a difference across wider areas of need, creating a national network of support. These groups, the big, quietly important organizations and the smaller, more personal foundations, team up to create a wide network of care that really honors those who have passed away by helping the people who are still here.

Fisher House Foundation

Fisher House Foundation has facilities near military and VA hospitals, except for one near Dover Air Force Base that helps the families of the fallen. They offer a "home away from home" for military families during the most stressful and vulnerable times, when a service member or veteran needs specialized medical care far from home. A Fisher House isn't a fancy hotel; it's a safe place to feel stable in the scary and unpredictable world of a medical crisis. Their impact isn't just in big news stories, but in the countless nights of deep family relief they provide, helping healing happen without the heavy burden of travel and accommodation costs. Their service is a gentle, ongoing act of kindness that helps families stay together when they're most at risk of breaking apart.

Gold Star Wives of America

Gold Star Wives of America also shows how powerful it can be to share experiences. This strong group of people offers personal support and important advocacy for widows and widowers. They know that no one can truly understand the unique and devastating challenges of military loss unless they've been through it themselves. Their strength isn't loud; it's the quiet power of shared experiences and helping each other up. Through understanding and supporting each other, they make sure that those who face this most unique kind of loss never feel completely alone.

Gary Sinise Foundation

People like Gary Sinise show us that a country really appreciates its people by how much it cares in private, not just with big public shows. It's about the deep, ongoing support for those who've faced the toughest challenges, making sure they know they're not alone when they need it most. It's about helping families not just get through tough times but also grow and succeed, turning the silence of loss into a powerful sound of strength and the hope of a happy future, always remembering those who gave everything.

The Gary Sinise Foundation (GSF) is more than just a celebrity charity; it's a real, practical commitment to those who've served, showing that gratitude should be real and last a lifetime.

The GSF was started with the idea of "never forget," and it's all about helping active-duty military personnel, veterans, first responders, and their families. Even though the foundation helps a lot of people, it really makes a difference by focusing on what they need most, like boosting morale, supporting families, and making sure they can live independently.

The R.I.S.E. (Restoring Independence Supporting Empowerment) program is the heart of our mission, shining brightly as a testament to it. This program is all about building custom smart homes for America's most severely wounded veterans—those who've suffered catastrophic injuries, often leading to paralysis or the loss of multiple limbs. These aren't just any homes; they're digital command centers carefully designed to meet their specific daily needs, helping them regain autonomy and dignity. For a hero in a wheelchair, a smart home designed for accessibility turns survival into thriving.

But the GSF goes beyond just physical rebuilding. It's about ensuring the whole military family heals holistically. Initiatives like the **Snowball Express** help the children of fallen military heroes connect, heal, and find joy through peer support and annual all-expenses-paid trips to Disney World. This shows that the sacrifice of service affects the entire family, so we need support that addresses emotional well-being and the bonds of community.

The foundation also boosts morale by offering performances, dinners, and events that remind service members, both here and overseas, that their work is truly appreciated.

TAPS (Tragedy Assistance Program for Survivors)

The world feels upside down after a loss, and grief feels like a wild storm, organizations like TAPS (Tragedy Assistance Program for Survivors) are there to offer a comforting, steady hand. TAPS knows that while sorrow is deeply personal, it doesn't have to feel like you're alone. They go straight to the heart of despair, creating a safe space where you can finally acknowledge the raw, overwhelming pain with people who truly understand.

What makes TAPS so special is how they create "shared silences." These aren't just quiet moments, but sacred spaces where the unspoken grief of one heart echoes in another. A glance, a shared tear, or a quiet nod can say more than a thousand words that might not be quite right. In these carefully designed environments—through counseling, strong peer support, and gentle remembrance ceremonies—survivors start to realize they're not alone. Their individual experiences find echoes in the hearts of others who have walked a similar, painful path. This shared empathy creates a powerful, invisible support system, helping the raw edges of pain soften, feel less alien, and less overwhelming. TAPS teaches us something simple but revolutionary: while grief is personal, healing and learning to live again are deeply communal. They give survivors the tools, professional counseling, and valuable peer networks they need to navigate the complex journey from immediate shock to lasting memory, turning the isolation of individual sorrow into the shared strength of a compassionate community.

Even with the incredible emotional support from organizations like TAPS, the practical challenges of loss can be overwhelming for families who are left behind. Grief isn't just a feeling; it touches every part of life. When a service member dies, the family's financial and logistical security can be completely broken in an instant. The Gold Star family needs quiet moments to grieve and space to heal, but they also worry about keeping their family together financially dealing with increasing bills, uncertain futures, and the real risk of losing the home that holds so many cherished memories. The constant worry about money can make it hard to focus on healing, diverting the survivor's energy from processing their emotions to just trying to get by.

Tunnel to Towers Foundation

Tunnel to Towers Foundation (T2T) really shines, offering a strong and stable support system that addresses a basic need that is essential for all healing. T2T became well-known for its dedication to giving Gold Star families and catastrophically injured veterans mortgage-free homes. Mrs. Slutman's heartfelt statement that Tunnel to Towers paid off the mortgage on their home is a powerful example of how much relief this support can provide. Chris, a fallen hero, left behind not only a grieving family but also the lingering question of how to stay stable. By taking away the huge financial burden of a mortgage,

T2T doesn't just give a house; it gives a safe haven. It frees the family to focus their energy on rebuilding a life based on love and lasting memories, instead of being constantly worried about the tough logistics of just getting by.

This kind of support is important because it gives a safe, steady place to be. It helps families start to dream about and live a full life after losing someone they loved. T2T teaches us that being stable is the first step to healing. Without worrying about money, people can really grieve, remember, and find peace in the middle of all the sadness. The home, which used to make them anxious, becomes a safe haven, letting them focus on healing without the extra stress of money.

Even though TAPS and Tunnel to Towers have different ways of helping, they both want the same thing: to help grief turn into lasting memories and a strong, happy life. TAPS gives us the comfort of being with others who understand, where we can open up without feeling judged, and where the heavy weight of sorrow is lifted by everyone sharing their feelings. T2T offers the quiet, safe feeling of a home, a place where families can find peace and focus on healing without the added worry of money. Together, these organizations show us that true support is all about taking care of both the heart and the mind. They know that a grieving heart needs not only love and support but also a solid foundation to rebuild on. They also know that finding comfort in shared silence means understanding between people and the peace of a stable home. They don't take away the pain of losing someone—that's not possible—but they give us the tools, connections, and help we need to deal with our grief, remember our loved ones, and finally live a life full of meaning, strength, and the lasting power of love.

In a world that often shouts and shows off, the most lasting ways we remember people are usually hidden from the public eye. These are the quiet efforts of families like the Gold Star families, who, after facing unimaginable struggles, decide not to stay silent but to turn their deep pain into real, positive change. Even though their heroes are gone, their stories are not just memories in stone; they are lively and active acts of kindness. These are the roots that give hope.

The world often doesn't hear the quiet cries of someone who's hurt. For those who have loved and lost a service member, the everyday noise—traffic, chatter, small worries—can become too much to handle. When a military loved one dies, a deep silence falls on the survivor, but it's not a peaceful silence; it's a hollow echo. This mix of loud outside sounds and inside emptiness makes the survivor feel completely alone, like time has stopped, leaving them stuck in a painful present while everyone else moves on.

Robert Hendriks Foundation

To honor Robert's final sacrifice, we need to understand that the pain of losing someone also needs action that is just as strong. The memorial fund is dedicated to helping our local veterans' organizations to continue the meaningful work they do to honor and help to improve the lives of our American veterans.

Benjamin Hines Memorial Foundation

The Ben Hines Memorial Foundation continues his spirit of service by providing scholarships for the children our fallen heroes, as well as the children of our veterans and law enforcement officers, and by supporting their families in times of need.

Everyone's grief is different, and there's no one-size-fits-all approach. Here are some helpful suggestions:

Professional Guidance: Reaching Out to Mental Health Professionals

One of the first steps in healing is often talking to a mental health professional. These experts have the knowledge and tools to help untangle the complex feelings of trauma.

Evidence-Based Therapies

Therapies like Cognitive Behavioral Therapy (CBT), Eye Movement Desensitization and Reprocessing (EMDR), and Prolonged Exposure (PE) are specifically designed to help process traumatic memories, challenge negative thoughts, and gradually bring people back into life. A good therapist can create a safe space for individuals to work through their past at their own pace, giving them the tools to handle triggers and rebuild their sense of security.

Medication Management

For some, medications prescribed by a psychiatrist can help manage acute symptoms like severe anxiety, depression, or sleep problems. These medications can create a solid foundation for therapy, making it easier for healing to start. This decision is always made carefully with a medical professional.

Finding the Right Help

If you're looking for a professional, try to find licensed therapists, psychologists, or psychiatrists who specialize in trauma. You can start by checking online directories, asking your primary care doctor for recommendations, or reaching out to local mental health organizations.

The Importance of Support: Your support network is super important when you're healing. Having people around you who understand and care can really help you feel better and make it easier to get through tough times.

Even though professional help is important, connecting with others and feeling understood can also be helpful for your emotional well-being.

Talking to Trusted People

Sharing your journey with trusted family and friends, even in small ways, can really help you feel less alone. If you can gently explain PTSD to them, they can better understand and support you without judging you, which can create a safe space for you to be yourself.

Joining Peer Support Groups

Connecting with others who have been through similar things can be really validating. Peer support groups, whether online or in person, give you a special place to share, listen, and know that you're not alone. The empathy and wisdom you get from people who really understand can be incredibly empowering. Organizations like the National Alliance on Mental Illness (NAMI) or local community mental health centers often organize these groups.

Finding Community and Spiritual Support: For some people, finding comfort and purpose in a faith community or local volunteer groups can give them a sense of belonging and meaning, which can help them get back into a supportive social network.

Taking Care of Yourself: Self-Care and Coping Strategies

Healing isn't just about dealing with the trauma directly. It's also about building resilience, finding inner strength, and making time for peace and quiet in your daily life.

When flashbacks or anxiety pop up, try deep breathing, focusing on your five senses (sight, sound, touch, smell, taste), or guided meditation. These can help you stay grounded in the present, giving you a sense of security when your mind feels a bit off.

Physical activity and nature are also great. Gentle exercise, regular walks, yoga, or just spending time in nature can help release physical tension, improve sleep, and boost your mood. It's a wonderful way to reconnect with your body in a caring way.

Creative expression can be a wonderful outlet for many. Art, writing, music, or journaling can help you process your emotions, share your experiences, and turn pain into something positive. These creative activities allow you to express things that words might not be able to capture.

Healthy habits, like getting enough sleep, eating well, and having predictable routines, can give you a sense of control and stability, especially when things feel a bit chaotic after trauma.

In our digital world, there are lots of supportive resources available online and on mobile devices. They can provide you with immediate information and tools to help you through.

Reputable Websites

Organizations like the National Center for PTSD (U.S. Department of Veterans Affairs), the National Alliance on Mental Illness (NAMI), and the Substance Abuse and Mental Health Services Administration (SAMHSA) provide valuable information, self-help tools, and directories to help you find local support.

Mobile Apps

Apps such as 'PTSD Coach' (developed by the VA), 'Calm,' or 'Headspace' offer guided meditations, coping strategies, and tracking tools that you can use anytime, anywhere. Many of these apps are designed to help you manage your emotions and provide immediate support during tough times.

Crisis Hotlines

Knowing that help is just a call away during moments of intense distress can be a real lifeline. Keep national crisis hotlines or local mental health emergency numbers handy. (e.g., 988 Suicide & Crisis Lifeline in the United States.)

K9s For Warriors

K9s F Warriors provides highly trained Service Dogs to Military Veterans suffering from PTSD, Traumatic brain injury and/or military sexual trauma. With many dogs being rescues, the innovative program allows the K9/Warrior team to build an unwavering bond that facilitates their collective healing and recovery.

Paralyzed Veterans of America

PVA is dedicated to improving the lives of veterans living with spinal cord injuries and other paralysis-related conditions. Comprised of members who have experienced similar challenges, PVA offers a wide range of services, including advocacy for equitable healthcare access, assistance with navigating the Department of Veterans Affairs benefits system, and employment support programs to foster independence. The organization also empowers veterans through community-building initiatives, adaptive sports, and peer support networks, connecting individuals with experienced volunteers who provide guidance and encouragement.

Others

There are other services that are provided to Military Veterans and Gold Star families and too numerous to list all of them.

A Gentle Reminder on Your Journey

Just a friendly reminder that healing from PTSD is a deeply personal journey and often doesn't follow a straight path. There will be both good days and challenging days, and both are important parts of getting better.

Patience and Self-Compassion: Treat yourself with kindness. You wouldn't expect a physical wound to heal overnight, and emotional wounds need even more gentle care and time.

Finding Your Fit: What works for one person might not work for another. Be open to trying different therapies, groups, and coping strategies until you find what really works for you.

It is Acceptable to Seek Assistance: Reaching out is a sign of strength, not weakness.

You have an amazing capacity for healing, growth, and finding peace again. Use the resources you have, lean on those who care, and trust in your incredible resilience. A calmer and brighter future is waiting for you.

Chapter Fourteen

My closing Interview with Corbin

The Language of Scars and Stars

The silent echo of the phone hanging up brought an end to hours of exhausting honesty. But if the room was silent, my mind was a tempest. It was not a storm of chaos, but of memories and voices, an overwhelming convergence of all the interviews that had paved a path to this story. Each conversation, whether an almost silent whisper across a Zoom or discussions of his narration in a veteran's support group, had been a delicate excavation. We had dug through layers of time and grief, unearthing pain, unwavering strength, and the indelible scars of sacrifice. Now, at this poignant close, a complex set of emotions enveloped me, weighted and profound.

As I reviewed the mountain of notes, the terrible question formed: How could mere words, no matter how carefully chosen, how intricately arranged, ever adequately convey the depth of their sacrifice or mend the fractured pieces of lives irrevocably changed? The scale of the suffering, both seen and unseen, felt too vast, too deeply ingrained for any single narrative to truly address.

I was a storyteller, dealing in language, structure, and compelling narrative arcs. But sacrifice existed beyond narrative; grief existed beyond metaphor. The attempt to distill years of pain into a cohesive, digestible book felt reductive, almost disrespectful. The helplessness was the paralyzing fear that I would fail him; that my attempts to illuminate his pain would only render it flat, turning three-dimensional tragedy into a two-dimensional statistic. The reality of the quiet veteran battling PTSD, the constant vigilance and terror that lingered long after the enemy was gone, was a deep, silent war happening in millions of homes. How do you transcribe the weight of a memory or the sudden tremor of an anxiety attack? The gap between the experience Corbin lived and the story I could tell felt immense, a gulf I could not bridge.

This helplessness, however, did not lead to despair. Instead, it struck a flint against my soul, igniting a fierce, undeniable charge.

The paralysis broke, yielding to desire to do more than just report; it was a profound realization that transcended the journalistic assignment. This story was not merely a feature to be published in a book; it had become a personal impression etched into my very consciousness.

The impression was dual-pronged, focusing on both those who paid the final price and those who returned to fight silent battles.

The quiet room no longer felt heavy with sorrow, but vibrant with purpose. The silence that followed my discussion with Corbin was not an ending, but a deep breath taken before a marathon. The notes spread across my desk no longer represented a finished story, but a journey of sorrow and pride. The story of scars and stars was complete, but the weaving of its future, the work of honoring its truth, had just begun.

It is important to advocate for the Gold Star families, those who must navigate living ordinary lives while carrying an extraordinary, permanent void. The quiet ache of an empty chair at the dinner table that no amount of time or new faces could ever truly disguise. The devastating realization that the world had moved on, sometimes carelessly, while their world remained fundamentally shattered, tethered to a moment in time they could never undo. The sorrow was pervasive because I understood that these sacrifices were not theoretical; they were children who would grow up without a father's hug, wive who would spend decades speaking to photographs, and parents whose deepest fear was that the world would forget the remarkable spirit of the soul they lost. It was a grief that demanded respect, and it threatened to drown the objectivity I was trained to maintain.

Intertwined with this crushing grief was a brilliant, almost blinding thread of pride.

This was not the superficial pride of victory or fanfare, but an immense, humbling pride derived from witnessing the extraordinary human capacity for courage and commitment. It was pride in the Marines themselves; those who ran toward danger so others might walk away from it, those who committed their youth and strength to ideals far larger than their own lives. Their commitment, often misunderstood by the civilian world, was a profound testament to duty, honor, and loyalty which must not be relegated to a footnote in history books or an annual holiday mention. Their honor demanded not just remembrance, but tangible, ongoing support and understanding. We must amplify their voices, ensuring that the resources available – medical, educational, and emotional – match the magnitude of their loss. Their dignity must be protected, and their enduring love must be celebrated, not just pitied.

But my pride extended beyond the fallen to the resilient spirit of the families who survived them. They were the unsung heroes of this narrative, bearing their profound loss with astounding grace and dignity. They harnessed their pain, not to become bitter, but to become advocates, building foundations, supporting comrades, and ensuring the legacies of their loved ones remained vibrant and active. Observing their strength in the face of an impossible burden filled me with an awe that bordered on reverence. This pride was the necessary ballast, preventing the ship of my emotions from capsizing in the sea of sorrow. It affirmed that their story, however painful, was fundamentally one of light and profound human goodness.

We must champion the cause of Service members battling the silent, insidious war of PTSD and other invisible wounds. Corbin's quiet resolve, even amidst his lingering pain, demanded action. We owe them more than gratitude; we owe them systemic empathy. They must receive the understanding, the destigmatization, and the unwavering clinical care they deserve to heal and rebuild their lives. Their courage did not end on the battlefield; it continues every day they choose to fight for their future, and we must stand beside them in that fight.

Corbin's story is a powerful testament to the unbreakable covenant shared by those who wear the uniform. It speaks not only to the explosive violence they faced abroad but to the quiet, agonizing psychological toll that follows them home. He exemplifies the painful truth that service involves not just the willingness to fight, but the necessary courage to live and carry the memories of those who cannot. The bonds forged under fire are not severed by death; they are deepened and consecrated by the survivor's enduring devotion.

The impulse to connect, to share, to bear one another's burdens, is deeply ingrained in the human spirit. Let us never forget the stories of countless service men and women who gave their lives for our freedom, those returning with scars, and the Gold Star families who will forever feel the pain of a loved one giving the ultimate sacrifice.

Chapter Fifteen

Thank You

Flags for Heros

As I was writing Echoes Of Afghanistan, I was humbled by the local organizations within my own community and others that are dedicated to supporting veterans who served in the various Armed Forces. It's stories like theirs and those like Corbin's that are vital to continue remembrance of the sacrifice made so that we can be free.

To SSgt Corbin Sipe and Katrina Sipe for their desire to share their painful story to encourage others is truly commendable because it requires immense courage to transform private agony into a public lesson. They willingly sacrifice their own comfort, reopening old, traumatic wounds not for attention, but with the selfless intent of illuminating a path for people currently trapped in similar darkness. This act of profound vulnerability is a powerful testament to the human spirit, demonstrating that the deepest suffering can be processed and converted into a practical source of empathy and resilience for those who desperately need to hear that survival, and ultimate healing, is possible.

Local support groups, particularly Military Clubs, serve as essential pillars of support, providing a crucial and safe environment for active service members and veterans to share deeply personal experiences and offer vital mutual aid. These community spaces foster profound solidarity, ensuring that fellow Service members receive necessary emotional and practical support from those who truly understand the unique challenges and sacrifices of military life. I personally want to extend my deepest appreciation to the Heritage Shores Military Club for their tireless dedication to this cause, evidenced by their frequent fundraising initiatives benefiting veteran support organizations. Their commitment to public acknowledgment through displays such as the moving Flags for Heroes tributes displayed during various holidays, and monthly meetings, keeps the contributions of service members front and center in the community, ensuring their sacrifices are always remembered.

The Quilts of Valor Foundation, through their dedicated stitching and heartfelt presentations, offers a powerful and deeply personal way to honor veterans who have bravely served in combat on foreign soil. Witnessing these vibrant, handcrafted quilts, each a testament to countless hours of love and remembrance, being wrapped around our servicemen and women is an experience that resonates profoundly. It's a tangible embrace of gratitude, a symbol of thanks for sacrifices made far from home. In those moments, as the fabric settles, a quiet awe descends, a shared understanding of valor and service that often leaves the audience, and more importantly, the veteran recipient, touched beyond words, with not a dry eye in sight.

Countless other organizations across the country, such as the Veterans of Foreign Wars (VFW).

The unparalleled freedoms we cherish daily are not simply inherited rights but are meticulously protected and often paid for with immense sacrifice by the brave men and women serving in our armed forces. These dedicated individuals willingly step forward, leaving behind the comforts of home, family, and personal aspirations, to stand on the front lines against threats both foreign and domestic. They endure rigorous training, face unimaginable dangers, and sometimes make the ultimate sacrifice, all to safeguard the principles of liberty, justice, and peace that define our nation. Their unwavering commitment and selfless devotion ensure that we, as a society, can live secure in our homes, express our thoughts freely, and pursue our dreams under the banner of a protected democracy.

May We NEVER FORGET!

Chapter Sixteen

Definitions

Family of Vehicles

MRAP is a family of vehicles designed to provide increased crew protection and vehicle survivability against current battlefield threats, such as IEDs, mines, and small arms. The DoD initiated the MRAP program in response to an urgent operational need to meet multi-Service ground vehicle requirements. MRAP vehicles provide improved vehicle and crew survivability over the High Mobility Multi-Purpose Wheeled Vehicle (HMMWV) and are employed by units in current combat operations in the execution of missions previously executed with the HMMWV. **Credit DOD Programs: Mine Resistant Ambush Protected (MRAP)** The Georgia Defense Readiness Program – Training (GDRP-T)

The Georgia Defense Readiness Program – Training (GDRP-T) was a bilateral United States-Georgian training program launched on May 1, 2018. U.S. Army Europe supported the Georgian Ministry of Defense's training of nine Georgian battalions through 2021. The Georgian Armed Forces maintained a high-level of training for its international missions, and the GDRP training mission complemented and added to Georgia's interoperability as well as strengthened the country's territorial defense capabilities. U.S. and Georgian forces served alongside each other in Afghanistan and around the world, so it's fitting that we support them in this training effort to promote peace and European security. **Credit: Defense Media Activity – WEB.mil**

www.ingramcontent.com/pod-product-compliance
Lightning Source LLC
Chambersburg PA
CBHW071328150726
47997CB00002B/632